PLANETS OF ROCK AND ICE

PLANETS OF ROCK AND ICE

From Mercury to the Moons of Saturn

Clark R. Chapman

Charles Scribner's Sons
New York

Dedicated to my daughter,
Cinette Ariana Chapman,
whose insatiable curiosity
exemplifies what this book
is all about,
and for letting me write it
without interruption (most of the time)

Library of Congress Cataloging in Publication Data
Chapman, Clark R.
Planets of rock and ice.

Rev. ed. of: The inner planets. c1977.
Includes index.
1. Planets. I. Title.
QB601.C46 1982 558.9 82–692
ISBN 0–684–17484–7 AACR2

This book published simultaneously in the United States of America and in Canada— Copyright under the Berne Convention.

1 3 5 7 9 11 13 15 17 19 F/C 20 18 16 14 12 10 8 6 4 2

A revised and expanded edition of The Inner Planets.

Printed in the United States of America.

Contents

*Illustrations 1–25 follow page 62;
illustrations 26–47 follow page 128.*

Some Important Lunar and Planetary Space Missions vii

Preface xi

1. Planetologists 1
2. Craters: Planetary Chronometers 13
3. Uniformitarianism and Catastrophism 27
4. Fragments from the Solar System's Birth 43
5. An Inner Planet Revealed 63
6. The Inside View 76
7. The Vapors of Venus and Other Gassy Envelopes 89
8. The Moon: What Did We Learn from Apollo? 111
9. Galileo's Worlds of Ice, Rock, and Sulfur 129

10. Saturn Encounter: Resplendent Rings, Exotic Moons 148

11. Mars: Changing Perspectives 164

12. The Earth in Its Planetary Context 184

13. The Galileo Project and the Future of Planetary Science 196

Index 213

Some Important Lunar and Planetary Space Missions

SPACECRAFT	NATION	LAUNCHED	TARGET	ARRIVED	DESCRIPTION
Mariner 2	U.S.A.	8/62	Venus	12/62	Measured temperature, magnetic field
Ranger 7	U.S.A.	7/64	Moon	7/64	First close-up photographs, before impact
Mariner 4	U.S.A.	11/64	Mars	7/65	22 close-up photographs
Venera 3	U.S.S.R.	11/65	Venus	3/66	First impact on another planet
Luna 9	U.S.S.R.	1/66	Moon	2/66	First lunar soft landing
Surveyor 1	U.S.A.	5/66	Moon	6/66	First U.S. soft landing on moon
Lunar Orbiter 1	U.S.A.	8/66	Moon	8/66	First of a series of lunar orbiters to study potential Apollo landing sites

Some Important Lunar and Planetary Space Missions *(Continued)*

SPACECRAFT	NATION	LAUNCHED	TARGET	ARRIVED	DESCRIPTION
Venera 4	U.S.S.R.	6/67	Venus	10/67	Made measurements during descent through atmosphere
Mariner 5	U.S.A.	6/67	Venus	10/67	Remote measurements of atmosphere
Surveyor 5	U.S.A.	9/67	Moon	9/67	Measured chemical composition of lunar surface
Mariner 6	U.S.A.	2/69	Mars	7/69	Martian flyby
Mariner 7	U.S.A.	3/69	Mars	8/69	Companion spacecraft to Mariner 6
Apollo 11	U.S.A.	7/69	Moon	7/69	First manned landing, brought back lunar rocks
Apollo 12	U.S.A.	11/69	Moon	11/69	Second manned landing
Luna 16	U.S.S.R.	9/70	Moon	9/70	Unmanned landing, brought back lunar rocks
Apollo 14	U.S.A.	1/71	Moon	2/71	Manned landing in Fra Mauro region
Mariner 9	U.S.A.	5/71	Mars	11/71	Relayed pictures and data from Mars orbit for about a year
Apollo 15	U.S.A.	7/71	Moon	8/71	Manned mission to Hadley Rille

SPACECRAFT	NATION	LAUNCHED	TARGET	ARRIVED	DESCRIPTION
Luna 20	U.S.S.R.	2/72	Moon	2/72	Unmanned; sample brought back
Pioneer 10	U.S.A.	3/72	Jupiter	12/73	Traversed asteroid belt, studied Jupiter and its environment
Venera 8	U.S.S.R.	3/72	Venus	7/72	Transmitted data from surface of Venus for 50 minutes
Apollo 16	U.S.A.	4/72	Moon	4/72	First manned landing in lunar uplands
Apollo 17	U.S.A.	12/72	Moon	12/72	Final manned lunar mission
Pioneer 11	U.S.A.	4/73	Jupiter Saturn	12/74 9/79	First man-made object on an escape trajectory from the solar system
Mars 4 and 5	U.S.S.R.	7/73	Mars	1/74	Mars orbiters
Mariner 10	U.S.A.	11/73	Venus Mercury	2/74 3/74	First close-up photographs of Venus and Mercury
Venera 9 and 10	U.S.S.R.	6/75	Venus	10/75	First photographs from surface of Venus
Viking 1	U.S.A.	8/75	Mars	6/76	First measurements from surface of Mars
Viking 2	U.S.A.	9/75	Mars	8/76	Companion mission to Viking 1
Luna 24	U.S.S.R.	8/76	Moon	8/76	Returned two-meter-long soil core

Some Important Lunar and Planetary Space Missions *(Continued)*

SPACECRAFT	NATION	LAUNCHED	TARGET	ARRIVED	DESCRIPTION
Voyager 1	U.S.A.	8/77	Jupiter	3/79	Reconnaissance of giant planets and their satellites
			Saturn	4/80	
Voyager 2	U.S.A.	9/77	Jupiter	7/79	Companion to Voyager 1; will reach Uranus in 1986, then go to Neptune
			Saturn	8/81	
Pioneer Venus	U.S.A.	5/78	Venus	12/78	Specialized orbiter spacecraft
		8/78	Venus	12/78	Multiple atmospheric entry probes
Venera 11 and 12	U.S.S.R.	9/78	Venus	12/78	Probes descended through atmosphere
Venera 13 and 14	U.S.S.R.	10/81	Venus	3/82	Latest of Soviet Venus explorers

Preface

The final week of August 1981 marked the end of a beginning. A silvery robot, packed with microminiaturized electronics, completed the mission designed by its builders back on Earth. Its tiny motors had worked imperfectly, yet it had managed to transmit a precious stream of data back home about the stupendous gas-globe Saturn, its encompassing rings, and its menagerie of satellites and moonlets. Years later, with luck, Voyager 2's instruments might measure other, dimmer worlds on the outskirts of the solar system. Voyager 2 was just one in a series of remarkable devices designed to learn about the cosmos and to pave the way for the human beings who would someday venture forth from the planet of their birth. But now it was forgotten by politicians of a once-spacefaring nation: no more spacecraft were following behind to expand on its exploration of Jupiter and Saturn.

Inexplicably, the golden age of planetary exploration has been put on hold. A science advisor to the president actually claimed recently that American science would benefit from budget cuts; he opined that attempts to understand our planet and its neighbors were not fundamental or necessary. While shortsighted American officials lose sight of mankind's destiny to explore space, new planetary missions are underway in Europe and Japan, and the Soviet Union continues its ambitious Venus program. So the human species, if not the nation that first put a man on another world, will continue to explore space.

The scientists, engineers, and journalists who worked during those August days to share Voyager 2's revelations with the public could not believe or accept the White House policy decision. Nor was it accepted by the thousands upon thousands of high school students and just plain folk who came to Pasadena, California—the center of mission activity—to attend the Planetary Festival, sponsored by the newly formed Planetary Society of Carl Sagan and Bruce Murray. Many people wore buttons proclaiming "On to Uranus!" or "Halley's Comet is Coming!" But the grim reality was that Voyager 2 wasn't designed to last to Uranus (though it may) and too little time remained to mount an American mission to meet Halley's inexorable arrival in early 1986.

So, in August 1981, the American people sat back in their living room chairs to soak up the infectious excitement of the final planetary encounter. Saturn's remarkable rings shimmered with incessant motion. The small moon called Enceladus exhibited geological terrains as complex—and recent!—as planets a thousand times its mass. With startling clarity, Voyager 2's lenses captured the intricate, many-hued clouds caught in Saturn's racing jet-streams. Project scientists such as Ed Stone and Brad Smith narrated their first impressions of the new pictures and data while newscasters such as ABC's Jules Bergman reflected on the irony of this last "picture show."

The final Saturn encounter was certainly fine entertainment. Together with successful flight tests of NASA's Space Shuttle, Voyager's achievement was one of the few triumphs in a year of assassinations, deepening recession, and a crackdown in Poland. Yet planetary exploration is an enterprise of far greater long-term significance to our culture than the fleeting experiences and discoveries of Encounter Week. It epitomizes and synthesizes the elements of basic scientific research—an enterprise that has helped create civilization as we know it and that beckons us to deal with the future.

Many journalists and politicians, not trained in science, don't realize that Encounter Week was just the tip of the iceberg of Voyager 2's contributions to human knowledge. Indeed, superficial "discoveries" can be glimpsed from a picture, leading the presidential science advisor to call planetary exploration mere "show biz." But scientists must then analyze the data, compare with theory and experiment, and integrate other planetary data. The interdisciplinary synthesis may take years or decades in order to fully capitalize on the investment in Voyager. But, shockingly, we may not do it. No new funds have been appropriated for studying the vast storehouse of data about Jupiter and Saturn, which Voyagers 1 and 2 have added to that obtained about the moon, Mars, and other planets a few years earlier. Instead, NASA's budget for data analysis has been cut back during each of the last few years.

Perhaps my chief purpose in writing *Planets of Rock and Ice* has been to show how thoroughly fascinating and important the planets are when we examine them closely and with more insight than we can by just looking at TV pictures during encounter. Seen through the eyes of scientists, after years of analysis and creative thought, the planets play a central role in understanding our own world, the processes that make it habitable for us, and the circumstances that have made intelligent life possible at all.

The scientific method is unfamiliar to most people, but the concepts of science are easily grasped: they are logical and relevant. In this volume, I try to circumvent the barrier of jargon that discourages people from grappling more deeply with science. By the end of the book, you will know more about the planets. More important, you will appreciate the uncertainties that surround our new "knowledge" and the fundamental issues that compel us to continue exploring the solar system. Finally, you will comprehend a bit about scientists, and about why scientific ideas can never be separated from the perceptions and emotions of the men and women engaged in creative research.

Planetary science is a youthful discipline. Only one university has a full department of planetary science and just a handful offer students the chance to earn a degree in the field. But I believe the developing methodology of planetary science will be a model for many other natural sciences in years to come. The narrowing specialization of many sciences is increasingly at odds with our need to deal with the complex, interrelated forces that shape our environment. Planetary research, however, is almost uniquely interdisciplinary: specialists and generalists from backgrounds as diverse as physics, biology, chemistry, and geology all struggle together to understand complex, distant worlds. I am confident that this approach and the "planetary perspective" will have profound, unpredictable effects on the more down-to-earth sciences.

Let me cite one example of how planetary research is abetting a fundamental reexamination of the tenets of fields that might seem totally removed from celestial influence: paleontology and evolutionary biology. Superficially, it is the topic of "the asteroid and the dinosaur": the thesis that an extraterrestrial body crashed to Earth and wiped out the dinosaurs 65 million years ago. As with other issues in science, this idea is less simple than that but it is of central importance.

In the early nineteenth century, scientists divided the history of life, as recorded by fossils, into four periods. Abrupt changes in the fossil record that marked the end of an era and the beginning of the next were first interpreted as "mass extinctions" of plants and animals due to catastrophic disasters such as floods. Later there developed a Darwinian synthesis of biological and geological evolution, based on slow, gradual change. Since then, there has never been a thorough reconciliation of the fossil extinctions with Darwinian theory. The recent rise of "creation science" is one misguided result; more sober, rational revisions of Darwin's theory now underway are others.

Recently a striking new hypothesis was proposed to explain the extinctions at the end of the Cretaceous—the final phase of the Mesozoic era. Nobel Prize-winning physicist Luis Alvarez, his son Walter, and several others discovered unusually great amounts of the rare trace element iridium in a fossil clay layer from Italy at the exact end of the Cretaceous (the so-called K-T boundary). Soon other scientists, using instrumental techniques perfected in the analysis of moon rocks, found iridium enriched in K-T boundary samples from around the world. In one location, the iridium-rich layer was found—to within a precision of just fifty years—to have been deposited *exactly* when many species of forams suddenly became extinct, 65 million years ago. (Forams are tiny planktonic animals, whose shells accumulate in sea-floor sediments.) Most of the Earth's original endowment of iridium now lies deep within the core, but iridium is abundant in asteroids and other heavenly bodies. The Alvarez team proposed, therefore, that an asteroid had impacted the Earth, filling the air with its iridium-rich debris, blotting out sunlight for months, and causing the extinctions of plant and animal life associated with the K-T boundary.

A memorable meeting, held in Utah in October 1981, brought paleobiologists and planetary scientists together for the first time. Although they didn't doubt the iridium anomalies, most paleontologists were skeptical of the "far-out" Alvarez hypothesis. Once treated to a heavy dose of planetary perspective, however, many changed their minds. First, experts on asteroids and comets explained how often the Earth is struck. Planetary geologists presented overwhelming evidence for past lunar and planetary bombardment. Physicists described computerized attempts to "model" what happens when an asteroid hits a planet's surface. Next, paleontologists listened to a NASA specialist on planetary atmospheres describe how long the iridium-rich dust might remain suspended in the air to dim the sun and change the climate. Finally, a planetary chemist showed how an asteroid might destroy the Earth's ozone layer and poison the oceans.

Later, one dinosaur specialist told me that he used to think the notion that Arizona's Meteor Crater was due to an impact was advertising hype for a tourist trap. He assured me that he would think about dinosaur bones differently in the future. But the interdisciplinary education worked both ways. The physical scientists and planetary geologists came to appreciate difficulties in linking extinctions of plankton with the demise of dinosaurs. Dinosaur bones, it seems, are rare and difficult to date. There is conflicting evidence about the simultaneity of iridium anomalies and dinosaur extinctions, so many paleontologists regard the linkage as unproved or even unlikely. (Several attending a 1982 meeting expressed such bitter resentment of the Alvarez team "using" their beloved dinosaurs misleadingly to gain publicity that Walter Alvarez later vowed he would never again speak to the press about the subject!) I don't know if an asteroid sealed the fate of dinosaurs, but the issue is deeply relevant to the origin of the human species: some scientists think the disappearance of dinosaurs opened the ecological niches that permitted mammals to proliferate on Earth.

As participants left the Utah conference, there was little doubt that a planetary view of predictable catastrophisms would profoundly affect future thinking about the evolution of life on Earth. Conference organizer Lee Silver, of the California Institute of Technology, summed up the interactions this way: two evolutionary sciences, planetary and terrestrial, had converged on the realization that paleoecological systems might, indeed, respond sensitively to big projectiles flying through space. Silver described the developing revolution in planetary science as affecting the earth sciences as profoundly as the revolution in plate tectonics and drifting continents did two decades earlier. The Utah meeting was held many years after the flybys that opened planetologists' eyes to the pervasiveness of planetary cratering. But, according to Silver, scientific synthesis takes time and "it is only now becoming apparent that what we have learned from studying the planets will have a direct bearing on our understanding of the evolution of Earth."

Planets of Rock and Ice is much more than a revision of my earlier book, *The Inner Planets.* Since then there have been the Viking studies of Mars, the Pioneer explorations of Venus, and the remarkable Voyager expeditions to Jupiter and Saturn. Hence I have added four completely new chapters. The other chapters have been rewritten or revised. Thus the book provides a 1982 perspective on what we have learned about the planets during the first two decades of the Space Age. Sadly, the hiatus in the American space program ensures that the book will be up to date for years to come.

In the earlier book, I concentrated on the rocky, terrestrial bodies, from the innermost planet, Mercury, out to the asteroid belt. I now extend my focus beyond the belt to embrace the rocky and icy bodies studied by the Voyagers: the moons of Jupiter and Saturn and the countless moonlets that compose the Saturnian rings. Voyager data are less digested about giant Jupiter and Saturn themselves, and their more distant relatives, Uranus and Neptune, are yet unexplored. But I couldn't treat the interrelationships of planets without bringing up the outer planets from time to time. Still, my subject is mainly the smaller, solid planets and satellites and their relatively thin atmospheres. Maybe after a decade of studying Voyager's rich harvest of data, after Voyager's forthcoming visit to Uranus, and after Galileo has spent two years orbiting Jupiter, it will be time to look afresh at the outer planets.

I am grateful to several of my colleagues, especially Mike Malin, for reading several new chapters; others provided some illustrations. Kathy Heintzelman at Scribners has been especially helpful. My wife and daughter have been patient. Finally, I wish to acknowledge that the pictures and results described in this book were paid for by the American public: of every $100 in federal taxes, nearly $1 goes to NASA. Of that, 4¢ has supported planetary exploration. I hope that, after reading this book, you will agree with me that your investment in fundamental knowledge about the solar system was worth it.

Clark R. Chapman
January 1982

1

Planetologists

Even today, there are moments when what I do seems to me like an improbable, if unusually pleasant, dream: To be involved in the exploration of Venus, Mars, Jupiter, and Saturn. . . .

—*Carl Sagan, 1973*

Why are we fascinated by the planets? Perhaps it is because they are remote, exotic worlds, with vistas and climates almost unimaginable to Earthlings. We have traveled there vicariously with the astronauts, who brought back to the rest of us a new perspective on our world. To Earthbound beings, the existence and relative proximity of the other planets provide the same potential respite and escape from our busy world that the ocean does for traffic-bound Bostonians or the mountains ringing Los Angeles do for those immersed in the smoggy basin below. Though we rarely travel to the oceans or mountains, we are comforted by knowing that we can. So too for the planets: they are real places to visit.

The planets are also a potential economic resource. Our ancestors transformed the wilderness—they harvested forests, harnessed rivers, and mined the earth. They built railroads through valleys, mountains, and desert floors. They farmed and settled rugged, virgin land. The economic and technological progress has made us (some of us) the wealthiest, healthiest, and most educated people on Earth. The solar system is also a wilderness to conquer, for good or ill. Although direct economic returns, such as mining the asteroids for metals, are not imminent, there are indirect benefits already. Studies of the atmospheres of Venus, Mars, and Jupiter have helped us understand our own so that we can better predict the weather. The need to fit sophisticated instruments into small spacecraft has spurred miniaturization technology. And, of course, the Space Program has provided employment to tens of thousands of workers.

What we know of the planets has been achieved by the research of scientists: astronomers, geologists, meteorologists, and others. What motivates them to do their research? Why are they fascinated by the planets? The planetologists who study the solar system share some of the same magical-mystery-tour and economic-political interests in space exploration that stimulate public interest. But their chief motivations are ones peculiar to their profession as scientists. Their innate curiosity about the universe is satisfied by their own special mix of rigorous methodology and hit-or-miss approaches to answering the questions and problems posed by what they see.

Planetologists have all of the psychological and sociological motivations that lead people to pursue their life's endeavors: recognition by colleagues, money to feed a family, or just resignation to following the path of least resistance through life. Some would-be scientists were told they were good in science in high school, followed their counselors' or parents' advice, took science curricula in college, and eventually emerged into the world equipped to be nothing but scientists, however their personal motivations might have evolved in the interim. It was unfortunate if society's educational system disgorged them into an economy surfeited with scientists, or if the inducements of government-financed scholarships led them to study space science, or particle physics, only for them to find at the end of their scholastic careers that society had now become interested in ecologists, cancer specialists, and exploration geophysicists.

But beyond psychological and societal motivations, scientists have unique intellectual drives that differ even among the various disciplines. Perhaps what distinguishes scientists most from other educated men and women is their fascination with the solution of *simple* problems. It is a reflection of how poorly our schools teach science—and our culture disseminates it—that the common perception of scientists is that they deal with the most complicated and intricate problems, far beyond the ken of nonscientists. It may seem that scientific matters are beyond our everyday experience, yet human affairs are extraordinarily complex by comparison. It is particularly true that the public regards physicists' work as the most abstruse kind, while actually physicists study some of the simplest things in nature. For instance, the physicist who specializes in celestial mechanics despairs if the number of point-masses in his study approaches the number of fingers on one hand. The work of the geologist or biologist seems a little less remote, but actually the formation of a single valley, or the biology of a single organism, is far more complicated than anything studied by physicists. And sociologists, who study aggregates of the most complex animals on Earth, are the researchers most "understood" by laymen, although their problems might seem better left to the poets and philosophers.

The physical scientist (physicist, chemist, astronomer) is fascinated by a simple problem because there is a chance that something about it can be understood in a fundamental way. But few things in our everyday lives are that simple (physicists learned why a ball rolls downhill some while back). The physicist's studies may ultimately transform human society, but the physical entities he or she hopes to understand in minute and complete detail are frequently so microscopic and specific that there is little connection with familiar things. We all live because electrons surround atomic nuclei, but we can't hold either in our hands. As lay people we can vaguely empathize with biologists and cages of cute little mice. Many of us are ourselves amateur psychologists. But the world of the physicist's simple playthings seems remote, mysterious, and a little frightening.

Consider this ordering of intellectual disciplines: mathematics and physics, astronomy and chemistry, meteorology and oceanography, biology and geology, anthropology and psychology, economics and sociology, history, literature, and the arts. The motivations of the practitioners differ across this spectrum, as do the public's comprehension and the relevance it sees for personal lives and society. The list ranges from subjects that are simple to those that are complex; from thorough and fundamental to partial and preliminary understanding; from a highly logical to an intuitive and imaginative approach; from objects of study far removed from our everyday experience to our own environment and ourselves; from matters with only indirect social relevance to the very guts of our interpersonal relationships and our perceptions of the world.

As we traverse this intellectual spectrum we move also from a high degree of predictability to none at all: the physicist's prediction that water will flow downhill tomorrow never goes awry; the astronomer is pretty good at predicting when the sun will rise, although predictions of the splendor of Comet ("of-the-century") Kohoutek were a bit optimistic; the meteorologist's weather forecasts are a trifle less trustworthy; the prognostications of government economists are in the same league as the foretellings of palm readers; and the disciplines at the end of the list do not even claim predictive ability.

Still another criterion establishes the intellectual spectrum and is fundamental to the motivations of the practitioners: the information content of the data. It cannot be simplicity of subject that places astronomers, in whose purview are countless planets and stars, before the geologists, who study a single world. But the geologist is inundated with data: visual observations of hills, stream beds, fault scarps, rock beds, fossils, ripple marks, and—at still smaller scales—the texture, color, and fabric of the minerals that compose the rocks. The geologist can also measure the hardness of rocks, chemical composition, radioactivity, and so on.

Such data are collected from locations around the world; the combined information is enormous and only the smallest portion of it can be digested. While the geologist can hardly understand the detailed history of a locale—every trickle of water, every vibration, and the crystallization of each mineral crystal—he or she can hope to decipher the largest and most important events. The geologist is thus one up on the social scientist in that the subject is not constantly changing as it is being studied.

In contrast, the astronomer has far fewer data to work with. While other worlds may be as complex as Earth, the information that can be obtained about them is minimal. Until this past decade we could not touch, hammer at, or experiment with samples of the stars or larger planets. All we knew about them was what astronomers could see from a distant vantage point beneath our blurry atmosphere. (Modern technology does permit astronomers to "see" planets' radiation of infrared and radio wavelengths, beyond the colors seen by our eyes. Rockets and satellites send instruments above the atmospheric molecules that absorb and block the ultraviolet and other wavelengths.)

The other worlds and stars are so far away that they appear very small; most seem to be mere points of light even as magnified in the greatest telescopes. One can measure the brightness of a point of light at every spectral wavelength, how that brightness changes with time, and the position and subsequent motion of the point of light. That is absolutely all the data a ground-based astronomer can obtain from which to learn the nature of an object. Through intelligent interpretation and synthesis of such limited data from other similar points of light, the astronomer can apply scientific laws and other observations and can draw conclusions about the object that might seem wonderfully detailed to the nonastronomer. So the astronomer, although not approaching the geologist's daily complexities, will have learned absolutely everything possible about this object from the data on hand. If the astronomer designs a better instrument, uses a bigger telescope, or spends more time measuring the object through the telescope, a more accurate record of the object's brightness may be obtained. But the data are always limited, so the astronomer extracts every possible clue, striving to reach that ultimate limit to his or her understanding of the object set by the meager means available.

An astronomer thrills at successfully solving the well-contained problem by rigorously stretching the few data to the limit. But a geologist is intrigued by the intellectual feat of recognizing the particular elements from an overwhelming mass of data that will lead to a slightly better understanding of a complex process. Each scientist has motivations, abilities, and intellectual habits suited to his or her own discipline. Through the pluralistic efforts of all, the understanding of our environment—the universe—proceeds on many levels.

The simpler a problem, the closer we come to a fundamental solution. Conversely, the more we learn about a problem, the more complex it turns out to be, and the more difficult the solution is to attain. At times it almost seems that we are on a treadmill: better instruments yield more data; these raise discrepancies and complexities with respect to the simpler models that had been formulated from earlier data; and finally the data become so numerous that there is no hope of assimilating and understanding them all.

This problem is particularly acute in the planetary sciences. Until the last two decades, our information about the major planets was obtained solely from telescopic observations of their small disks. Even the sharpest telescopic photographs of the planets were much inferior to a picture of someone's face taken with an inexpensive box camera. The fuzzy image of Mars through a telescope can be thought of as an array of several hundred separate points of light. By comparison, a television image, whether on your home set or a spacecraft vidicon picture, is an array of more than 100,000 separate picture elements. Recent technology has brought an explosion of planetary data: all nonredundant data obtained about Mars before 1965 could have been printed in a small book, but Mariner 9, which orbited Mars for a year in 1971 and 1972, sent back more data about Mars to Earth *each hour* than had been collected previously in all human history.

Now volumes of Viking pictures form a small library, and this does not include the tapes of data returned by other non-imaging instruments on the Viking orbiters and landers.

The information explosion has been even greater in the case of the moon, which has been photographed and mapped in even greater detail than Mars. Moreover, the astronauts brought back half a ton of moon rocks which will be analyzed in microscopic detail for decades to come. It is said that a sufficiently large quantitative change is a qualitative change. So it is for planetology; accustomed to "milking dry" the limited available data, astronomers now feel inundated by spacecraft data and have turned the study of the moon, Mars, and Mercury over to geologists. Several years ago an astronomer friend told me, "Mars is dead." He wasn't speaking of the results of the Viking Mars-lander biology experiments, which searched for life on Mars. Although he was premature, he meant that there was nothing further that astronomers could learn about Mars. Astronomers are also turning Venus, Jupiter, and Saturn over to meteorologists now that vast amounts of atmospheric data have been returned by Pioneer and Voyager missions to those planets.

This scientific version of musical chairs has been evident in the programs of the annual meetings of the Division for Planetary Sciences of the American Astronomical Society (AAS/DPS), an organization of the several hundred astronomers whose major interest is studying the planets and solar system. In 1971 and 1972, 36 percent of the scientific talks were concerned with the moon and Mars. By 1976 only two talks dealt with the moon and fewer than 9 percent with Mars. Meanwhile, papers concerning the asteroids and the natural satellites of the planets increased from 9 percent of the program in 1971 to 37 percent by 1976. Why satellites and asteroids? They were barely more than points of light in a telescope and, except for the moons of Mars, had yet to be examined closely by any spacecraft. What could be a more ideal pursuit for a planetary astronomer than to turn a telescope toward those small, faraway worlds and discover all that could be learned about them? Despite the exciting Viking and Pioneer findings about Mars and Venus in the late 1970s, the planetary astronomers' meetings in the 1980s remain dominated by small bodies and the giant outer planets—all fertile ground for telescopic observation.

Planetary scientists are guided, beyond their intellectual preferences, by spiritual and experiential motivations. Although astronomers spend most of their time in laboratories or classrooms, they make periodic pilgrimages to mountaintop observatories. Beneath these cold, distant pinnacles on the doorstep of the heavens are spread the Earth's desert wildernesses. Civilization is manifested only by the headlights of an infrequent automobile and the glowing auras of cities on the far horizon. Amid the towering pines an occasional dome is silhouetted against the stars and the Milky Way. Inside, the giant machinery looms skyward, slowly revolving to follow the stars. The astronomer sits below, monitoring the instruments, checking the guiding, and gazing at a slice of the heavens through the open slit in the dome. Drowsily intoxicated by the rarefied air and the shock of the body's circadian rhythm adjusting to nighttime wakefulness, the astronomer is receptive to the immanence of Nature.

I recall once at the National Observatory: the night was still and silent save for a few howling coyotes. I was trying to measure the light from some faint asteroids, a task that grew more difficult as the full moon rose to dominate the sky. But sitting on the carpeted floor looking out the slit, I noticed the moon growing dimmer. Its left side began to be eaten away by (I then realized) the shadow of the Earth. Finally it became a faintly glowing ocher orb, and the scattered moonlight in my photometric signal vanished entirely. Between swallowing my midnight lunch and marveling at what was to me a quite unexpected total lunar eclipse, I hurriedly measured several asteroids right next to the ghostly moon in the sky. As dawn approached, the moon emerged from the Earth's shadow and grew full again. As it sank toward the west, I was finishing measurements in the brightening eastern sky. As the sun's first rays peeked over the Rincon Mountains 70 miles to the east, I trudged down from the dome to sleep as the rest of the world awakened. I knew that the magnetic tape was now filled with good data on tiny worlds never before measured. I was overcome by a profound sense of satisfaction and I felt as renewed as the moon had been earlier that morning.

A geologist laces up thick-soled boots, packs away a compass and trusty rockhammer, rents a four-wheel-drive Jeep, and heads off to wildernesses as strange as those on other worlds. The exhilaration is tangible as the geologist shades the sun burning down from cloudless skies and climbs a barchan sand dune in a desert that hasn't seen a drop of rain in over a century. Another geologist clambers over the plateaus and valleys of frigid Antarctica, where liquid water, which nourishes most of the world, hasn't existed for millennia. Perhaps this remote spot is more like the surface of Mars than anywhere else on our variegated planet.

Still other geologists seek planetary analogs in rare places where the Earth is being born. They step gingerly across glassy, solidified lava that was molten but a week before. Their next step might pierce a fragile bubble and plunge them into a fiery cave below. The mists and sulfuric fumes swirl over the unearthly new landscape, threatening to suffocate the explorers; yet they know that the deadly gases are the Earth's freshest emanations and they are reminded of how "unnatural" is the oxygen-rich air we breathe. The intrepid explorers watch a glowing pink tongue of molten rock ooze past them down a glassy channel. Tired and hungry, they stop to bake a frozen pizza on a freshly crusted lake of liquid rock. But the molten pudding shifts, carrying the simmering meal a few feet out "to sea." They plunge onward still hungry, for one does not dip one's toes into a puddle a thousand degrees hot. At last they stare down into the fiery hellhole they have journeyed to see—a caldera a quarter of a mile across. Silvery-orange liquid rock from the Earth's deep interior swirls in turbulent, incessant motion. At any moment the precipitous cliff from which they observe in awe may break loose and crash into the all-consuming liquid furnace below. Undaunted and dedicated, they set up their equipment before retreating to relative safety. Volcanologists' lives are as fraught with danger as those of any explorers of centuries past; and from their measurements and observations some further understanding may be gleaned of processes that shaped the Earth, and Mars, and other worlds too, eons ago.

The geologist in the wilderness epitomizes man against nature. A tiny, transitory creature, whose life will be over in a mere second of geologic time, attempts to discern the life processes of Earth, whose metabolic rate is a million times slower than its student's. Hammering at the rocks pregnant with data, the geologist is in physical contact and combat with the Earth, like a mosquito attacking Goliath. Ever so grudgingly, the Earth yields its life history.

It is evening. The planetary physicist's mind is alert and he can concentrate without interruption. He is grappling with the origin of Saturn's moons. His tools are a sharpened pencil and a pad of paper. He has a long-practiced facility for manipulating formulas that rigorously describe the laws of nature. For years he has wondered about the numbers that define these moons' orbits; they form fractions too exact to be mere coincidence. Recently a spacecraft obtained a new datum about Enceladus, one of Saturn's moons. And pictures showed the moons as real, physical worlds to be understood. Newly motivated, the physicist constructs logical relationships between the moons' behavior and principles established by Newton, Maxwell, and Chandrasekhar. Using Greek letters for otherwise wordy concepts and unknown quantities, he organizes the elements of the problem that he hopes to synthesize by simplifying the equations.

He had intended to study geophysics in graduate school, to merge his childhood talent in mathematics with his collegiate fascination with the International Geophysical Year. However, the Space Age arrived and his interest was attracted to the orbital mechanics of artificial satellites launched to study the Earth. For his thesis topic he had ventured farther afield to analyze forces affecting comets and planetary satellites. This evening he is again in familiar territory. He realizes he could apply an old trick to simplify the equations.

After an hour and three pages of penciled notes, he has a

shorter formula. But it is peculiar; it violates his intuition. Rechecking his arithmetic, he finds an erroneous "+" instead of "−" in front of one expression. A cup of instant coffee settles his frustration and he plods through the algebra once more. The eventual result is less simple but more elegant. The exponents of several terms pose another impediment, though.

The physicist would rather have solved the whole problem alone, but he is practical enough to call for needed help. He swivels in his chair and switches on the gray computer terminal in the corner of his office. He types in some passwords, then a few lines of computer code that will direct the solution of his equation by a trial-and-error method. The straightforward calculation would have taken days for him to work himself. Now electrons flow for 20 seconds in the circuitry of a remote computer. Suddenly a swift green point of light dances across the screen. Measuring from the flickering graph, the physicist realizes his theory is compatible with the spacecraft evidence. Yet the curve bends sharply, unexpectedly downward to the right. That might imply unusual behavior for Mimas, another Saturnian moon. But the new puzzle can wait, for it is now 2:00 A.M. Human intellect has taken one further small step toward comprehending how planets evolve, and why.

In a windowless, air-conditioned room in the basement of a university building, another planetary scientist finishes preparing her experiment. She was trained as a chemist but now calls herself a meteoriticist. The elaborate machine before her cost over $100,000 and incorporates the latest advances of American engineering technology. For the past many months she has been fabricating auxiliary equipment and calibrating the instrument so that she can be sure the readings will be accurate. Now she has dialed in the proper numbers and is ready for her first measurement.

She picks up a stone that has been lying in a nearby tray, her hand trembling ever so slightly. This small lump of rock is older than any material known to mankind—billions of years. For the last many millions of years it has traveled through space to be delivered to Earth, and then to her laboratory. She returns it to the tray and picks up instead a small plate she constructed several days earlier; it contains a thin slice, only a few hundredths of a millimeter thick, from the same meteorite. She puts the plate in the sample chamber of the instrument, pushes a couple of buttons, and watches expectantly as the preprogrammed instructions are carried out. A fine beam of electrons plays across the thin section; from the reemitted X-rays, the precise chemical composition of each tiny spot on the slide will be measured.

Though she has never looked through a telescope or mapped a geological quadrangle, the meteoriticist regards herself as descended from the purest planetological tradition. For decades, even centuries, before the astronauts returned the first moon rocks, her colleagues and their progenitors had been studying rocks that came from planets a hundred times farther away than the moon. After all, it was the Nobel Prize–winning chemist Harold Urey who had written the definitive book on the planets long before modern astronomers thought our neighboring worlds worth studying. Soon the meteoriticist will have the microscopic chemical analyses that will permit her to test her theory about the processes that assembled this rock, grain by grain, an unfathomable eternity ago.

2

Craters: Planetary Chronometers

The code in which [terrestrial] planetary histories are recorded will be cracked by geologic mapping, for it is the spatial relationship of different bodies of rock that tells the sequence of events. . . . But the absolute age of the rocks is another problem: the correlation of geologic time on an interplanetary scale. . . . If the frequency of meteoroid impact on the different planets can be established, the ages of their surfaces can be estimated from the distribution of superimposed impact craters.
—*Eugene Shoemaker, Robert Hackman, and Richard Eggleton, 1962*

About 25,000 years ago a small band of native American hunters slowly made its way among the sparsely scattered pines then growing at the southern edge of the mesas that have been inhabited for the last few hundred years by the Hopi Indians. The hunters were descended from nomads who had trekked across the land bridge from Asia to exploit the riches of the vast continent. Spurred by the knowledge that they needed meat within a few days, the men were single-mindedly tracking the nearest prey.

A sudden flash of light in the sky above startled them. They turned and, shielding their eyes from the burning radiance, they glimpsed a brilliant ball of fire streaking toward the plains to the south. It exceeded the sun in splendor, and in a few short seconds it reached the horizon in a final blinding explosion of light. The fireworks were snuffed from view a few seconds later by a columnar pall of dust that lifted upward and outward from the point of impact.

Temporarily forgetting their quest for food, the men stood oblivious to the twittering birds nearby, gazing in amazement as the mushroom-shaped dust cloud towered ever more slowly upward from the horizon. There had been legends of fireworks and of spectacular clouds emanating from the black volcanic hills to the southwest, but this blinding bolt of fire from a cloudless sky was something they had never heard about, let alone witnessed, before.

Without warning the ground jolted and shook violently, knocking the hunters from their feet. Then the earthquake quickly subsided. But as the men struggled back to their feet they saw a wall of dusty haze expanding toward them across the plains below. They began to run away from the edge of the mesa. But before they had gotten far, there was an exploding thunderclap and their still world was filled with a roaring, rumbling noise. Terrified, they dropped again to the ground just as a mighty concussion of wind swept over them. They shut their eyes to the swirling dust. The wind died and the hunters stood up once more, glancing fearfully toward the south and wondering what further surprises were in store for them. But little was to be seen through the haze. The hunters anxiously queried each other about the intentions of their gods in bringing these phenomena upon them in such sudden succession. They debated about whether to call off the hunt for the day, but hunger compelled them to press on. Within an hour, their world was back to normal again, except for some lingering haze, and stalking the animals was once again uppermost in their minds.

What the members of this hunting party could never have realized was that they had witnessed one of the most spectacular and unusual of the geological processes that shape the face of Earth. Since the prehistoric formation of the Arizona Meteor Crater just described, no other human beings have witnessed the nearly instantaneous creation of such a large landform.

Meteor Crater, about 1 mile across, was formed by the impact, at a speed of more than 10 miles per second, of a huge nickel-iron meteorite or asteroid fragment about 100 feet across.

Rare though such cratering events are in human experience, impact cratering is actually the dominant geological process in the solar system. The "mountains" on the moon, Mercury, and Mars are mainly the raised rims of old craters. The same is true of many of the satellites of the outer planets and probably true of the asteroids as well. Indeed, cratering may have been the dominant geological process on our own planet during its first half-billion years of existence, just as continental drift, running water, sedimentation, and volcanism have dominated terrestrial geology in the most recent half-billion years, during which plant and animal life have proliferated on Earth.

As we compare our own countryside with the moonlike cratered surfaces of most of the other rocky and icy worlds in the solar system, we might ask why there are so few large craters on Earth that even a relatively tiny one only 1 mile across is a famous landmark. On the other hand, if interplanetary space is so empty, how can there be so many scars from impact explosions on most planetary surfaces?

The answers to both questions involve time. To be sure, interplanetary space is very empty indeed. If we were to reduce the scale of the inner solar system (those planets closer to the sun than Jupiter) to fit inside Houston's Astrodome, the Earth would be the size of a pea and there would be scattered throughout the rest of the stadium a few dozen microscopic dust particles capable of producing craters on our Earth-pea similar in scale to typical lunar craters, such as those shown in Illustration 9. The Earth-pea and a few dozen dust grains would be moving along their orbits in the Astrodome at an imperceptibly slow speed, similar to the speed of a watch's minute hand. Space is very empty, and the chances for collisions might seem nil. But after 4½ billion years, which studies of meteorites have shown is the age of the solar system, our Earth-pea would have traveled 2 billion miles in the Astrodome, and each dust grain a similar distance. After having covered such great distances within the confined volume of the Astrodome, the Earth-pea probably would have struck most if not all of the dust grains!

The Earth, then, probably has encountered most of the projectiles that have crossed its orbit in space. Others passed so close to the Earth that our planet's gravity altered their paths, sending them to wander in the outer solar system, perhaps ultimately to be ejected from the solar system altogether. Many of the bodies that failed to strike the Earth have scarred the surfaces of the moon and other planets. There are a lucky few asteroids in the inner solar system that by chance have missed the planets so far, while one by one their siblings died explosive deaths. Other asteroids now nearby were originally in safer orbits that did not cross the orbits of the larger planets, but they have shifted recently into Earth-crossing paths.

Where are all the craters that must have been formed on the Earth? Why don't we see them? Perhaps we do see a few of them, if we look carefully enough. Some geologists speculate that Hudson's Bay in Canada, or at least part of it, may be the flattened and flooded remnant of a huge asteroidal impact crater. The remains of several other probable craters, up to tens of miles across, have been discerned from aerial photographs by the trained eyes of photogeologists. The extremely eroded condition of those several large craters that have been found suggests why the many others are completely absent: compared with the immense durations between major impacts on Earth, the geological processes that deform and erode landforms proceed exceedingly rapidly.

During our own lifetime the modification of the landscape by floods, glaciers, and earthquakes has been very slight indeed. And the major forces in the Earth's crust that cause the continents to drift apart and crash together, creating whole mountain chains, occur even more slowly. Yet they occur rapidly enough to have erased all evidence of large terrestrial craters except in those ancient continental cores called pre-Cambrian shields. Most of the couple of hundred terrestrial craters that have been discovered are located in shields, such as the one that comprises much of Canada and some of the north central United States. Such regions have been protected from the mountain building, flooding, volcanism, and other destructive geological activity common along continental margins.

Meteor Crater itself is already appreciably eroded, although it was formed only a few tens of thousands of years ago. Although it is still impressive, dozens of arroyos and channels have been carved down its inside slopes by the raging runoffs of hundreds of thousands of thunderstorms. In a million years there may be little or no trace of Meteor Crater; yet a million years is only a few ten-thousandths of the age of the Earth. Given the near eternity of geological time, the crustal processes active on the Earth today can wipe the landscape virtually clean of craters.

Instead of asking why the Earth today has so few craters, we should wonder that the moon and most other planets have retained so many. The answer is simple: those planets must lack the powerful erosive processes active on our world. (Logic forces a second alternative upon us: perhaps the other planets share the Earth's geological activity but are being struck by thousands of times as many asteroids as the Earth. Although we have no assurance that the bombardment is the same on all planets, it would take a cosmic marksman of exceeding skill to single out the moon and other planets for bombardment but avoid hitting the Earth.)

One of the major conclusions of modern planetology is that our planet is, compared with most of our neighbors, a geologically active world. The familiar mountains, valleys, hills, and coastal plains have resulted from processes special to the Earth. For decades we had viewed our world as a typical planet orbiting a typical star. In 1965 scientists and laymen alike were shocked when Mariner 4 revealed a moonlike landscape on the planet Mars rather than the mountains and valleys to which we are accustomed. In the early 1970s, Mariner 10 disclosed Mercury to be yet another crater-scarred world. But just when we were beginning to regard moonscapes as the norm in the solar system and our own planet the solitary exception, Voyager 1 found a small world orbiting Jupiter that is in a state of geological agitation greatly exceeding even that of the Earth (see Chapter 9). And so our perspective on our own world is evolving. It may not be entirely coincidental that the scientific revolution taking place in geology, epitomized by new conceptions of continental drift and plate tectonics (see Chapter 12), happened simultaneously with the beginning of planetary exploration.

Craters or pits are conceptually and geometrically simple. Indeed, they are common, from the tops of frying pancakes to the old battlefields of Indochina. Whenever a sudden, outwardly radial force is exerted on a horizontal surface in a gravitational field, a crater is likely to result. The Earth's volcanic explosion craters (called maars) bear a striking resemblance to impact craters, and debate long raged over whether the large lunar craters were of volcanic or impact origin. Detailed studies of craters on the terrestrial planets (Mercury, Venus, Earth, the moon, and Mars) have revealed subtle diagnostic differences, and the debate on lunar crater origin is now largely resolved in favor of impact. But the major features of a crater (its size, depth, and distribution of ejected material) can be predicted chiefly from the energy imparted to the ground at a point. It matters little whether it is the kinetic energy due to a meteorite's mass and velocity or the explosive energy of a bomb: roughly the same size crater results each time from the sudden release of the same energy into rock or solid ice. In an instant, the explosive energy liquefies and vaporizes part of the rock and imparts a violent shock that expands into the ground, tearing rocks apart and heaving them upward and away. The larger, slower moving fragments build up on the exterior rim, while high-velocity fragments land at great distances from the crater.

Impact cratering experiments have been done at NASA's Ames Research Center near Palo Alto, California, using hypervelocity bullets traveling many miles per second. NASA experimenters have learned how crater size and form change with different velocities, bullet masses, angles of impact, and surface material strengths. Scientists also have a much better idea now of the interplanetary population of asteroids, comets, and smaller projectiles that have created craters in the past and still occasionally impact on planets. Using the 48-inch Schmidt camera-telescope on Mount Palomar, astronomers have surveyed the fainter asteroids, especially those passing relatively close to the Earth and

Mars. Theoreticians have calculated the permanence and longevity of these projectiles and the chances of their impacting a planet or being moved by the gravity of other planets into more distant orbits that cannot intersect the planets.

Thus scientists have a firm grasp on all major aspects of the cratering process. But understanding the cratering process is just the beginning, not the end, of planetologists' interests in craters. By applying this knowledge, scientists have been able to decipher the relative geological histories of the moon, Mars, and Mercury in the absence of fossils, which in the last century were necessary for establishing the Earth's chronology.

Little was known about the geological history of the Earth until the early 1800s, when Sir Charles Lyell and others noticed that different groups of fossils of extinct species occurred in an orderly sequence in deeper and deeper layers of stratified rocks. Although the concept of biological evolution was yet to be developed, geologists pragmatically recognized that if a layer of limestone contained certain ammonoid fossils, it must have formed during the Devonian Age (now known to have been about 350 to 400 million years ago), even though in all other respects that limestone was indistinguishable from limestones formed before and since. By tracing the different fossil distributions in rock outcrops throughout the countryside, European geologists developed a stratigraphic sequence of rock units ranging from ancient to modern. Later, fossil groups and rock units were correlated around the world and a coherent sequence of rock-forming and erosive periods was synthesized for the entire Earth.

A relative sequence of geological events provides no clue to their absolute ages. To say that coal deposits were formed in Pennsylvania at the same time coal was deposited in England, during the Permian Age, does not distinguish between the notion that the Permian occurred during one tropical summer in 2350 B.C. and our present belief that it began 280 million years ago and lasted for 50 million years. The relative history of terrestrial geology has been converted to an absolute one by the techniques of geochronology, perfected in the middle of this century. Certain radioactive elements found in rocks are known from physics to decay, in known durations, into different elements of special isotopic composition. From measurements of such isotopic components in a rock, its absolute age may be accurately determined. The oldest rocks dated on the Earth were formed over 3.8 billion years ago, which is a common age for most lunar rocks. The same geochronological techniques applied to meteorites generally show them to be about 4.6 billion years old.

For Mars, Mercury, and the other planets and satellites, we do not have rock samples to date. Nor can we even establish relative sequences from fossils. Here is where craters come to the rescue. Let me begin with an analogy.

Consider a postman (of the old-fashioned variety) walking along the sidewalk from house to house, delivering mail. Or shall I say trying to walk along the sidewalk, for let us imagine that it has been snowing for two days. It is not a heavy blizzard, but the snow has been mounting steadily and is a foot and a half deep on the undisturbed lawns. The postman is no stratigraphic geologist, but he can conclude when, in the past couple of days, the different families were out shoveling the snow from their walks. The snow is a foot and a half deep in front of the first house and no trace of the sidewalk is evident; either these people are on vacation, are sick, or are just plain lazy, our postman concludes, for clearly nobody has pushed a shovel across this walk for two days. In front of the next house the snow is only three inches deep—evidence that the walk was shoveled earlier that morning. Apparently the next sidewalk was last cleared the previous day, since about a foot of snow has fallen on it since. At the last house on the block, the sidewalk is practically bare, with only a few scattered snowflakes sticking on the cold cement; the postman now notices Mr. Smith still at work, finishing his driveway.

Substitute *lunar surface* for *sidewalk* and *cratering projectile* for *snowflake* and you understand the basis for the lunar relative stratigraphic sequence, first derived in the early 1960s by scientists at the Astrogeology Branch of the U.S. Geological Survey (U.S.G.S.). While all lunar regions are more heavily cratered than anywhere on Earth, some are more cratered than others. This is an important observation, for it implies (reasonably assuming that interplanetary projectiles strike all sides of the moon equally) that the moon has not been a geologically dead body, just passively recording the scar of each impact. Rather, on some parts of the moon, the craters once formed have been covered or otherwise erased by active geological processes, and only the more recent ones survive.

The provinces on the moon most devoid of large craters are the maria, or lunar "seas," the dark, circular patches visible with the unaided eye from Earth. Compared with the crater-upon-crater zones on the brighter uplands of the moon, the maria have only about $1/20$ as many large craters. Even among maria provinces, crater densities differ by up to a factor of 4. If we could be assured that the impact rate has been constant, we might conclude that the maria were formed only $1/20$ of the moon's age ago, or about 200 million years ago, when dinosaurs first roamed the Earth. But unlike our postman who witnessed the two days of steady snowfall prior to making his rounds, we cannot presume that the interplanetary bombardment has always been steady. More likely there was an early "blizzard," a phenomenon that would invalidate the 200-million-year-age estimate.

Observations other than those of crater density can be made from lunar photographs, taken both through telescopes and from spacecraft, which clarify relative sequences on the moon. For instance, specialists in photointerpretation might quickly recognize the scalloped feature in a photograph as the edge of a huge once-molten lava flow that spread across the mare surface, obliterating all preexisting craters. One need not count craters to conclude that the land underneath was formed before this latest flow lapped upon it! Only in small localities do such geometrical superposition relationships reveal relative age sequences. Crater count comparisons are required to establish the ages of these flow units relative to features elsewhere on the moon.

Applying such procedures to the best lunar photographs, geologists have devised a relative sequence of lunar geological history that is as reliable as that for the Earth. Craters, then, have served admirably in place of fossils. Yet before the Apollo 11 astronauts brought back lunar rocks in 1969, we had virtually no *absolute* age calibration for the moon. Still, the picture wasn't totally bleak. In the mid-1960s Arizona astronomer William Hartmann, among others, argued that the maria had to be much older than

200 million years. He estimated the rate at which the Earth, and thus presumably the moon as well, is bombarded by interplanetary debris. He used counts of the nearby asteroids, of the meteorites that fall each year, of the meteors that flash through the night sky (meteoroids too small to make it to the ground), and of the comets that pass by. By estimating the sizes of these projectiles, and how large the craters they might create would be, Hartmann showed that the present cratering rate over 4½ billion years would create only about the number of craters visible on the lunar maria, not the twenty-times-greater number on the highlands. Hartmann concluded that there must have been a blizzard of bombardment sometime, and he suggested that it most reasonably happened near the beginning of lunar history. Hence, long before the Apollo landings most lunar specialists believed the maria to be 3½ to 4 billion years old, and that the numerous highland craters formed earlier, during the final stages of formation of the moon itself.

Shortly before the landing of Apollo 11 some geochronologists, disinclined to believe that ages could be established by any method other than their own, dismissed the validity of the crater count method of determining ages. To make matters more confused, one influential astrogeologist then obtained some classified military data that showed (to the privileged few with security clearances) that many more objects were entering the Earth's upper atmosphere than had been estimated from meteor observations. But apparently these military data suffered from exaggeration, for when the geochronologists measured the ages of Apollo 11 lava basalts from sparsely cratered Mare Tranquillitatis, they turned out to have formed about 3½ billion years ago after all! The terrestrial bombardment rate today is admittedly still somewhat uncertain and could be higher than the average over the last few billion years. The whole question of bombardment rates is an important area of active current research. A large telescopic camera is being designed now for the express purpose of studying the population of interplanetary projectiles in the Earth-moon region.

I have described how the unprejudiced manner in which cratering impacts strike all parts of a planet equally enables scientists to develop relative geological histories for the planets. Beyond that, the remarkable similarity of the initial shapes of craters of all sizes enables planetologists to study the nature and styles of erosion and landform degradation on other planets.

The branch of geology known as geomorphology is concerned primarily with the processes that carve canyons, erode mountains, and otherwise shape our landscapes. Running water is the chief erosive agent on the Earth, although the chemical weathering of rocks, glacier movement, and many other forces help to shape it too. The moon is a simpler place, since water is absent and the sole erosive agent is the abrasion of surface rocks by impacting meteoroids. In times past, the destructive impacts of asteroids and the flooding of basins by molten lava played major roles, but those were the only forces.

Mars is a more complex world than the moon, although today it too lacks running water. Unlike the moon, Mars has a thin atmosphere, and the photographic record from the Mariner 9 spacecraft revealed abundant evidence of the abrasion and filling of Martian landforms by windblown dust and sand. In past ages, lava flooding and cratering changed the face of Mars, but so did running water and other processes. The shapes and forms of Martian craters, as they have been gradually obliterated over the eons, reveal much about these forces that have shaped the surface of the Red Planet.

To oversimplify, imagine a dusty Mars where craters of all sizes are being formed regularly, while dust is continually settling and gradually burying them. (Other erosive processes, singly or in combination, degrade craters in roughly similar ways, but dust deposition is easy to visualize and is certainly one of the processes currently active on Mars.) Each crater is initially formed as a fresh, bowl-shaped excavation like Meteor Crater in Arizona. But as the dust settles, the crater bowl becomes shallower and shallower until it is finally filled in. Of course, small craters are buried much more rapidly than larger ones, but there are always new small craters being formed while others created before are only partly gone. So at any time there is a complete range of craters of all shapes, but only up to a certain size.

Suppose the total depth of dust deposited on Mars since its

surface formed is 1 mile. Then many generations of small craters would have been covered and only the most recent ones would still be there. But craters larger than 20 miles across (with depths of 1 to 2 miles) would still be visible; even those formed very early in the history of Mars would not yet be completely buried. Thus, as the Mariner 4 spacecraft first revealed in 1965, the proportion of small to large craters on Mars shows a lack of small craters compared with the proportion of small to large lunar craters or small to large asteroids. The Estonian astronomer Ernst Öpik, who has resided for some decades in Northern Ireland, first correctly ascribed the absence of small craters on Mars to a major erosive process on that planet. (Öpik has always been one of the most resourceful planetary physicists. He still recounts the tale of his departure from Tashkent during the Bolshevik Revolution of 1917. The train had to stop at little stations for hours and even days as local stationmasters of uncertain loyalties debated what to do. But on more than one occasion Öpik pointed to the red or white glow of the aurora borealis in the northern skies, suggested it was the light of the appropriate advancing army, and got the train rolling again!)

Consider the shapes of Martian craters. They range from fresh craters to highly degraded ones. In the example described above, both cratering and dust deposition occur continuously, so for all sizes there is the same proportion of craters in each stage of degradation. There is an analogy with age distributions of human beings: we could classify people as children, young adults, middle-aged, and elderly. Although children are always being born and the elderly dying, the percentage of people in each of the four age groups remains constant, at least in a stable society.

Societies, however, are rarely stable. Age statistics of Americans reveal a disproportionate number of young adults (especially those in their early thirties). Even if we knew no history at all, we might infer from this demographic fact that there was a baby boom during the post-World War II years when these young adults were born. Rather analogous studies of the proportions of Martian craters of different classes reveal anomalies compared with the stable dust-filling example previously described. In the mid-1970s, Ken Jones (then of Brown University) and I concluded that there must have been an episode in Martian history when the crater obliteration rate greatly exceeded the cratering rate in comparison with more recent times.

It would be too simple to ascribe this episode of crater degradation to dust alone. But whatever combination of processes was active during this time, it had a major effect that is still visible on the Martian landscape. The relative stratigraphic sequence for Mars, which was established using the same principles I described earlier for the moon, shows that modification of cratered highland topography coincided with early volcanic activity on Mars and with the formation of a network of channels throughout the cratered provinces. This fascinating period of Martian renewal took place after the major cratering bombardment had slackened, but long before the greatest Martian volcanoes reached their present stature. It was probably between 2 and 4 billion years ago, but we cannot say exactly until we have some Martian rocks to date or until we establish the Martian cratering rate more precisely than we have so far.

What was happening on Mars during this episode of geological frenzy? Was it a period of climatic warmth, when the Martian atmosphere was temporarily Earth-like, and great wind and rainstorms eroded and filled the craters and carved the channels? Or was this the result of heat flowing out from the planet's interior, with accompanying volcanism, melting of permafrost ground, and a rising water table, causing the Martian highlands topography to be sapped from within and partly disintegrate? We won't know the answer until we can decide for sure just what geomorphological processes degraded the craters and whether or not the channels truly were carved by runoff from Martian rainstorms.

Perhaps the obliteration episode represents a synergistic combination of external and internal effects. The interior warmth of Mars peaked midway through the planet's geological history, perhaps due to radioactive heating or the even more dramatic process of core formation. Then enough frozen ground may have melted to outgas the first (and last?) substantial atmosphere Mars ever had. In a bootstrapping, feedback effect, the thickening air may have absorbed more sunlight and further warmed the climate. Mars may have reached the pinnacle of its planetary vigor with simultaneous global volcanism and atmospheric turmoil. Subsequently its internal radioactive engine decayed, its atmosphere froze out again at the poles and underground, and Mars became once again the dry, cold, practically airless world it is today. These enticing scenarios are hypotheses suggested by our analysis of Martian crater shapes, but we must explore the Red Planet in much more detail before choosing among them.

3

Uniformitarianism and Catastrophism

And the waters prevailed exceedingly upon the Earth; and all the high mountains under the whole Heaven were covered.

—*Genesis 7:19*

A wild animal is alarmed by instability in its ecological niche, a potential threat to its life. A sufficiently widespread calamity may threaten an entire species. Mankind has advanced by modifying the environment and so far has managed, more or less, to adapt. But the accelerating changes of the twentieth century have been psychologically disturbing—and for good reason. We discover that we have introduced carcinogens into our food, water, and air, while pollutants and the use of aerosol sprays threaten to plunge us into a sudden climatic change, perhaps a new ice age. Some people would promote large-scale strip-mining, continue dumping asbestos into Lake Superior, and relax pollution standards for the sake of our immediate economic health.

They believe that we can adapt limitlessly and that science and technology can rescue us from potential catastrophe. But it is a faith in technology that our instincts tell us may be poorly placed. Hence, the alternative impetus toward leading simpler, more natural lives.

The optimum environment of our dreams is the Garden of Eden or a tropical paradise, where the landscape is natural and fixed, the climate mild, and seasonal changes slight. The real world, with its storms, earthquakes, and floods, is not so constant. But as contrasted with our twentieth-century "improvements," we conceive of Nature as having been fairly constant over the millions of years during which mankind evolved. We feel comfortable with our "uniformitarian" conception of Nature's constancy and we fear change, especially catastrophic change.

Those who believed in an active Divinity and the wretched state of mankind and who looked to salvation in the next world were often inclined to believe in catastrophic changes in this world. No catastrophe was beyond the power of the Old Testament God to bring against His wayward children. Yet He might also save mankind, as He saved Noah from the Deluge. For one whose fate was in God's hands, no constancy of environment was required for salvation. Catastrophism is still proclaimed in "The World Tomorrow" broadcasts of the Worldwide Church of God, which predicted as late as the 1960s that unprecedented catastrophes would culminate in the early 1970s, prior to the divine salvation of the chosen few.

Since the uniformitarian/catastrophic dialectic is so fundamentally rooted in our psychology, culture, and religion, it is not surprising that the science most concerned with the natural environment—historical geology—has been greatly affected by philosophical and religious beliefs about environmental stability. Even debates concerning the geology of other planets are being waged today on the same philosophical battleground.

Most modern scientists assume the constancy of the laws of physics. The Book of Genesis, literally interpreted, and the theories of the catastrophist Immanuel Velikovsky are in conflict with even that assumption. But the uniformitarianism on which modern geology was founded nearly two centuries ago as a reaction to biblical catastrophism is a much more sweeping but less secure assumption. Increasingly, strict uniformitarianism seems incompatible with the scientific evidence, although some geology texts still rely on it and many geologists are still in its grip. It may well be the unfolding history of the solar system inferred from space exploration that will ultimately synthesize a reasonable blend of uniformitarianism and catastrophism, free from religious or philosophical bias.

I need not dwell on biblical catastrophism. If there were any truth to Archbishop Ussher's calculation of the date of Creation as 4004 B.C., the geological processes we witness today could hardly have carved deep canyons, laid down thousands of feet of sediments, and deposited aquatic fossils a thousand miles from the nearest sea. Even today, the visitor to the Grand Canyon is hard pressed to accept geologists' contentions that the river carved the whole chasm in only millions of years. By the early 1800s the Deluge of the Old Testament was deemed inadequate to explain European geography and mass extinctions apparent in the fossil record. So natural philosophers invoked a whole sequence of catastrophes, supplemented by divine repopulations of the world by ever more advanced animals.

Uniformitarianism, enunciated by James Hutton in the late eighteenth century, emerged from the Age of Reason, although its conceptual foundation can be traced from the medieval church and even the Greeks. It is the doctrine that "the present is the key to the past," that the Earth may be understood by assuming that the natural forces acting today have always been acting similarly in the past. Great canyons result from rocks being washed downstream grain by grain, and rock layers are built from sedimentation, grain by grain. Released from the 4004 B.C. constraint, the history of the Earth stretched back into the indefinite, unchanging past.

But catastrophism did not die easily, and by the early nineteenth century geologists had split into ideological camps. Uniformitarian geologists, led by Charles Lyell, hardened their hypothesis into dogma. It then took a while to gain their acceptance of even such limited catastrophes as Ice Ages. By the twentieth century it was axiomatic that the fixed continents were in perpetual balance between the steady forces that uplift and create new land and those that gradually wear it down. That Africa might once have been adjacent to South America was regarded by the prevailing uniformitarians as being as outlandish as the notion that craters on the Earth and moon were created by catastrophic collisions.

A methodological assumption akin to uniformitarianism pervades science: among alternative hypotheses, the simplest and most straightforward is to be preferred. If the Earth as we see it can be explained by present processes, why invoke catastrophes or changes? Of course, changes or catastrophes are not thereby disproven, and should contradictory evidence subsequently appear, a more adequate theory must be sought.

However, geologists read more into uniformitarianism than methodological simplicity. They increasingly felt that environmental constancy was a proven fact and that geological processes necessarily act within narrow limits. Rather than abandoning a simple model, geologists went to great efforts to explain contradictory evidence in uniformitarian terms. More than a century ago, William Whewell attacked Lyell's uniformitarianism, asserting, "Whether the causes of change do act uniformly;—whether they oscillate only within narrow limits;—whether their intensity in former times was nearly the same as it now is;—these are precisely the questions which we wish Science to answer. . . ." But geologists rarely cared to investigate whether the Earth might have been different in the past or to determine the degree to which geological processes have been nonuniform. After all, the whole superstructure of their science was built on uniformitarian assumptions; to ask a geologist to question them would be like asking a priest to be skeptical of God or an army general to question the utility of war.

The processes that shape our world are far from constant. From gentle breezes emerge tornadoes, packing winds of many hundreds of miles per hour. A mountain of cinders once grew in a Mexican cornfield. The largest lake in California, the Salton Sea, was formed practically overnight by diversion of the Colorado River. In 1883 most of the island of Krakatoa vanished in an explosion that threw as much as 20 cubic kilometers (5 cubic miles) of rock and dust into the sky. And the devastation done on a single spring day in 1980 by Mount St. Helens is familiar to everyone.

Though acknowledging that such minor catastrophes as floods, storms, and landslides "should not be discounted," the author of a recent geology text nonetheless concluded, "The cumulative effects of the constant downslope creep of soil, the gradual decay of rocks under the atmosphere, and the gradual removal of material by streams hour after hour, day in and day out, over millions of years are probably much greater than the effects of these minor catastrophes."

Recent research contradicts such a uniformitarian view, showing that the greatest changes occur during the biggest storms, earthquakes, or other "minor" catastrophes that take place during any interval. For instance, Eugene Shoemaker of the California Institute of Technology and the U.S. Geological Survey has shown that most change in the Grand Canyon results from severe storms and floods, not from the gradual action of the river and ordinary rains. Shoemaker retraced the route of John Wesley Powell's historic expedition down the Colorado River a century ago and took new pictures of sites photographed by Powell. One hundred years had left most of the canyon completely unchanged, Shoemaker discovered with surprise. But where change had happened, it was with a vengeance: catastrophic landslides had changed the whole vista!

It was the same Gene Shoemaker who, in his doctoral dissertation more than two decades ago, proved convincingly the catastrophic mode of formation of Meteor Crater. Astronauts who later walked on the moon, and thousands of others, have been privileged to be introduced to Meteor Crater by the energetic and jovial Shoemaker, who knows every pertinent outcrop. He estimates that during one in five of the hundreds of field trips he has led into the crater, someone finds a meteorite, a metallic fragment of the asteroid that hurtled into northern Arizona long ago. With infectious enthusiasm, Gene Shoemaker is glad to promote his future projects, including a sky-scanning telescope.

With it, he expects to find a near-Earth asteroid and launch men to lasso it, hammer at it, and understand it. Just such an object, Shoemaker and others believe, collided with the Earth 65 million years ago with such catastrophic force that the ensuing havoc—perhaps a several-month blackout of the sun, poisoning of the rain, and longer-term climate changes—wiped out the dinosaurs. Many other species of plant and animal life, on land and in the sea, succumbed in what was one of the greatest mass extinctions in the Earth's fossil record.

Modern research has continued to show, against dwindling uniformitarian resistance, that other powerful catastrophes have occurred, too. A vast region of valleys, basins, and buttes, 200 miles on a side, was created in eastern Washington State in just a few days 18,000 years ago, when 400 cubic miles of water from Lake Missoula suddenly broke through a glacial dam and poured across the land to the Pacific. There are indications that Ice Ages may begin almost instantaneously on a geological time scale, well within a human lifetime, so Thornton Wilder's image of the glacier advancing on George Antrobus's house may be nearer the truth than the allegory of all human history that Wilder intended in *The Skin of Our Teeth*. Great evolutionary changes have occurred also. For instance, about a hundred million years ago, the whole Indian subcontinent broke off from east Africa and rammed into southern Asia, creating the Himalayas. Although continents drift only inches per year, the idea that continents move at all was beyond the pale for uniformitarian geologists during the first two-thirds of the twentieth century (see Chapter 12). Other dramatic ways in which the planet Earth has evolved are still being uncovered; it is now also thought, for example, that our atmosphere was very different in the past.

It is most reasonable to constrain geological uniformity only by the laws of physics. Given sufficient energy, we are limited only by our imaginations in conceiving plausible ways in which the energy may be organized into geological activity. In making an inventory of accessible energy, we are ultimately limited by Einstein's $E = mc^2$, though such perfectly efficient conversion of mass to energy is not generally realizable. Until the present century the only known energy sources were sunlight (rather feeble for moving continents!) and the residual heat from creation (planets were assumed to be formed from molten rock). Lord Kelvin had shown a century ago that any primordial heat would be radiated away in 20 million years, so the apparent lack of adequate energy for propelling continents became an early obstacle to the acceptance of drift.

We now believe the major source of energy driving the Earth's active geology is heat released in the interior by radioactive decay of uranium, potassium, and thorium. Other sources of heat energy that may have been important when the Earth was young include gravitational (tidal) interaction among planets, explosive impacts and collisions, energy released when heavy metals plummet to form a central planetary core, interaction of a planet with the sun's "wind" of charged particles, and the decay of very short-lived radioactive elements.

While Earth scientists understand the generation of heat within our planet, it is much more difficult for them to understand how that energy is converted into the organized motions of the drifting continents and spreading sea floors that build mountains and shape our world. The complex suite of geological processes, belatedly recognized by geologists and woven into the new synthesis dubbed plate tectonics, is an awe-inspiring intellectual tapestry. It is much more intricate than the comparatively simple view that Mercury solidified and contracted around its immense iron core, causing its crust to buckle and form the large wrinkled cliffs photographed by Mariner 10 (see Illustration 4). Our own planet's solid but plastic mantle convects turbulently, in a manner similar to the rising and sinking of air near a thunderstorm or water in a pan approaching a boil. Our drifting continents are carried on the backs of the churning, subterranean currents, and molten rocks exude from cracks in the ocean floors. We will examine later how this fundamental internal life of our planet is ultimately responsible for the life-sustaining environments on the Earth's surface. How all this dynamic activity is produced is as yet poorly understood and awaits more detailed comparisons with the other rocky and icy worlds.

The laws of physics also permit such unexpectedly complex, organized motions as weather fronts, jet streams, and cyclones that are driven in our atmosphere by the energy difference of absorbed solar heat between the tropics and the poles. Still other planetary processes of almost unimaginable subtlety depend on chemical reactions, on temperature-dependent phase changes (e.g., water changing to ice), and on life itself. For instance, the rate at which our landscapes are eroded is limited not by the capacity of running water to carry sediments, but rather by the rate of rock "weathering": the chemical reactions with the air, water, and secretions of microscopic organisms that cause rock to become weak and susceptible to breaking and being washed away. Should we carelessly destroy the ozone layer and thereby do away with life on Earth, the atmosphere itself, which has been partly generated by life, might be altered. Weathering processes might thereby be modified, changing the familiar geological processes on our planet. This is speculation, but a planetary ecosystem is very intricate. How stable is the Earth's atmosphere, ocean, biology, and geology as a total system? We are still too firmly grounded in uniformitarianism to dare to find out.

Some inquisitive scientists have studied the stability of other planets, where we have fewer preconceptions and less to risk in discovering instability. Actually instabilities are common in nature, both the one-way kind (an avalanche) and the oscillatory kind (the cycle of ice ages). A more drastic form of climatic instability has been proposed for Mars by Cornell University astronomer Carl Sagan and his associates. Mars is now a cold, dry world with a tenuous atmosphere having a pressure that is only 2 percent of that at the top of Mount Everest. Yet Mars probably had huge amounts of water, other liquids, and gases when it formed. Sagan hypothesized that the polar ice caps on Mars might contain enough volatiles frozen out of the atmosphere (including water ice and dry ice) to constitute an Earth-like atmosphere if released. He calculated that any slight warming of Mars would begin a melting process that would eventually evaporate the ice caps and provide Mars with a thick atmosphere. Later the process might reverse. So Mars may have experienced one or more cycles of oscillation between its present state and an Earth-like state.

The effect of such climatological instability on the possibility for life on Mars is obvious, but the effect on Martian geology is equally significant. As described earlier, Mars shows evidence for a past episode of geological activity, perhaps including running rivers, which might be accounted for by the massive atmosphere that Sagan proposes. Other scientists doubt that sufficient volatiles are accessible to the atmosphere. But the correctness of Sagan's hypothesis is less important than the recognition it inspires that analogous catastrophic processes might well have occurred on other planets in times past, including our own. We certainly cannot assume past uniformity nor assume that other planets behave similarly to Earth.

The greatest catastrophes of all are those of colliding worlds. In one instant 4 billion years ago, more than half a billion years after the moon was created, an asteroid several dozen kilometers across struck the moon, forming the gigantic Mare Imbrium basin, which is bigger than Alaska. This moon-shattering explosion produced ridges and valleys a quarter of the way around the moon similar in size to the Appalachian ridges in Pennsylvania. Yet this event may have been only one of dozens of similar collisions and countless smaller ones that affected the inner planets during the so-called Great Cataclysm.

All catastrophes pale in comparison to the hypothesized Great Cataclysm. Whether it actually happened is not yet known, but eminent space scientists such as Gerald Wasserburg of the California Institute of Technology, George Wetherill of the Carnegie Institution, and Bruce Murray, director of the Jet Propulsion Laboratory, have proposed that nearly all of the great craters on the terrestrial planets were formed during one relatively brief episode lasting less than 1 percent of solar system history! Such a fantastic bombardment-blizzard in the inner solar system would also have cratered the Earth and Venus and shattered the asteroids.

This catastrophist's dream is by no means an implausible outgrowth of the processes thought to have formed the solar system. Cosmogonists now believe that many asteroids and small planets were left over after the major planets formed. For example, the accumulation of Uranus and Neptune was probably protracted, leaving remnant bodies for hundreds of millions of years after the inner planets had formed. Such a large, cometlike body might have been perturbed by the gravity fields of the giant outer planets into an elongated orbit penetrating the inner solar system, just as smaller comets are today. If during one orbit it passed near a planet such as the Earth, it would have been wrenched apart by great tidal forces—a scenario ten times more likely than that it actually impacted the planet. A remnant body from the accumulation zones of Earth and Venus might also have lasted several hundred million years before being tidally disrupted. George Wetherill, a man who developed his reputation in geochemistry and geochronology but whose expertise has expanded into celestial dynamics, believes that pieces of such a disrupted body might soon have struck and cratered all of the planets in the inner solar system.

Thus a Great Cataclysm could easily have occurred. One might conceivably occur again someday, with craters the size of Meteor Crater being formed somewhere on Earth every few decades. Is there evidence that such a catastrophe actually did happen? It is ironic, indeed, that some scientists' arguments that it did, have been imbued with uniformitarianism! The cataclysm was first proposed by Gerald Wasserburg and other geochronologists at the California Institute of Technology, whose laboratory they call the "Lunatic Asylum." By dating the first returned moon rocks, they found a sharp cutoff between plentiful rock ages younger than 4.0 billion years ago and the virtually absent older ages. Since many of the older moon rocks look as though they were once crushed and smashed, the "lunatics" ascribed the absence of older rocks to a "terminal lunar cataclysm"—an episode of saturation bombardment that destroyed almost all preexisting rocks.

Furthermore, while ages of moon rocks from several different sites differ only slightly, the accumulated numbers of craters in these areas—as counted from photographs—differ greatly. Proponents of the cataclysm argue that the lava flows that formed these surfaces occurred just before the end of the bombardment, so that regions only slightly older received many more impacts than regions formed a little later. Other scientists believe, instead, that the evidence favors a more gradual decline in the cratering rate during the final stage in accretion of the moonlets that formed the moon itself. They concede only minor statistical fluctuations in the cratering rate.

During the mid-1970s, controversy about the enormity and brevity of the lunar cratering bombardment 4 billion years ago spread to the evidence for a simultaneous bombardment of Mars, Mercury, and the moons of Jupiter. For these planets we have as yet no rocks to date, so how do we know when their craters were formed? Mars is everywhere less densely cratered than the lunar uplands, despite its proximity to the ostensible source of cratering projectiles—the asteroids. Couldn't the craters we see have formed continuously through Martian history, without any early bombardment?

Not according to Larry Soderblom, chief of the Astrogeology Branch of the U.S. Geological Survey in Flagstaff, Arizona. In December 1973 he gave an authoritative presentation before the first International Colloquium on Mars, held in Pasadena, California. He set out to "compare the lunar and Martian cratering histories by studying the rate of change of Martian impact flux implied in the geologic record provided by Mariner 9, assuming uniformitarianism in the geologic evolution of Mars." By assuming that land-forming processes on Mars had occurred roughly uniformly throughout time, Soderblom argued that the crude division of the Martian surface into heavily cratered landscapes and lightly cratered ones implied a sharply declining cratering rate near the beginning of Martian history.

Soderblom rarely used the word *uniformitarian* and has not vociferously advocated uniformitarianism. But he feels intuitively that the burden of proof that Martian geology has *not* been uniform rests with his critics. With one further uniformitarian jump in logic, the early cratering on Mars has been ascribed by several scientists to the same source as the early lunar cratering: if the lunar cratering was cataclysmic, it was on Mars as well.

What about Mercury? The Mariner 10 television team, led by Bruce Murray, adopted a working model for understanding Mercury's heavily and lightly cratered provinces that rests largely on analogy with the moon and Mars: ". . . We find it plausible to correlate the terminal bombardments on both the Moon and Mercury as resulting from a distinct episode that affected at least the Inner Solar System about 4 billion years ago. . . . But Mars also exhibits a heavily cratered surface. . . . The simplest and, to us, the most plausible explanation is that all three surfaces record the same episode of solar system bombardment."

In 1979 Voyager revealed that Jupiter's icy outer moons, Ganymede and Callisto, are also heavily cratered. The camera team, including its deputy leader Larry Soderblom, carried the terrestrial-planet analogy out to Jupiter: "The cratered terrains on Ganymede and Callisto certainly must date back to the late torrential bombardment period some 4 billion years ago, as do the cratered highlands of the moon." And the cataclysm even intruded into some analyses of the saturated crater fields found on some of Saturn's small, icy moons.

Not only has a Great Cataclysm been thought to be ubiquitous in the solar system, but there has been a uniformitarian presumption, as well, concerning planetary cratering *since* 4 billion years ago. That is crucial because, in the absence of datable rocks from planets and moons, the only way to establish an "interplanetary correlation of geologic time" is from cratering, using principles I described in the previous chapter. If we were to know, for instance, that there have been many more asteroids crashing onto Mars than onto the moon, then the fact that Mars has a crater density similar to that on the moon would imply Mars is a younger, geologically more active planet. The essential problem is to learn how the quantities and orbits of asteroids, comets, and other impactors have evolved since the putative cataclysm. The first step is to learn how many are out there today.

In the early 1970s, cratering chronologies for Mars and Mercury were derived not by studying asteroids and comets but rather by *assuming* that similar crater densities on their flatter, presumably postcataclysmic surfaces implied similar impact fluxes on those bodies. If the impact rates were the same, then any province on Mars or another planet cratered similarly to a province of known age on the moon would be deduced to have the same age as the lunar province. It seemed simple and straightforward to discount differing fluxes and differing chronologies, but was the uniformitarian assumption valid? To be sure, George Wetherill demonstrated that bodies in elongated, cometlike orbits, which cross all planetary orbits, have similar probabilities of striking each planet. Moreover, he showed that bodies orbiting near such moderately large planets as Earth and Venus are likely to have their orbits gravitationally altered by close encounters with Earth or Venus so that some of them would also impact Mercury and Mars. So a Great Cataclysm due to a swarm of Earth-Venus-region bodies would have affected all the inner planets. And a terminal bombardment due to comets might well be manifest throughout the solar system. Postcataclysmic cometary cratering might indeed yield similar cratering chronologies everywhere.

But until one has made a complete inventory of *all* possible impactors, a uniformitarian conclusion is suspect. Many known or plausible groups of projectiles are much less indiscriminate about their targets than are comets. Earth-Venus-region bodies and main-belt asteroids probably contribute few craters to Jupiter's moons and none at all to Saturn's. Perhaps Mercury's surface postdates any cataclysm and its craters are more recent, produced by a hypothesized swarm of intra-Mercurial asteroids, nearer the sun; such bodies could rarely venture out to Venus or beyond. Perhaps bodies from the inner asteroid belt, near Mars, have preferentially cratered the Red Planet. Gene Shoemaker has studied outer solar system populations of comets, many of which never make it into the inner solar system. Those comets that are "captured" into orbits nearer the sun lose their ices and shrink into smaller asteroids, or vanish altogether, before hitting the planets. Other comets interact primarily with Saturn and its satellites. In fact, some of Saturn's moons may have been cratered mainly by large ring particles and Saturnian moonlets, none of which would have escaped Saturn orbit to strike Jupiter's moons or the inner planets.

So while some projectiles, past and present, would have similarly cratered some or all of the planets, others would have had very specific targets. As the linkages between these populations and their reservoirs become understood, we will learn more about how widespread the proposed cataclysm was and about how uniform the more recent dribble of cratering has been. While the "uniformitarian" view of planetary impact-catastrophism apparently has some validity, clearly it is only a part of the story.

I would be remiss in discussing solar system catastrophism not to mention the late Immanuel Velikovsky. Inspired by a professional interest in psychoanalysis, he performed a solitary, unparalled synthesis of the implications of biblical and other ancient myths and legends for solar system history. Although based less on rigorous logic than on analogy and circumstantial evidence, his results were truly revolutionary. For instance, he proposed that Venus and Mars hurtled past the Earth in historical times, producing the devastating disasters recorded in legends from around the world. Velikovsky would have Venus launched from Jupiter as a monster comet shortly before it encroached on Earth, releasing manna from the heavens and causing the Earth to cease turning (among other feats), a phenomenon that would then account for the parting of the Red Sea for Moses and the Israelites.

His monograph *Worlds in Collision* was published in 1950, over objections of many astronomers who tried to censor his ideas, and was a best seller. Velikovsky was shunned by the scientific establishment for the rest of his life, though in the early 1970s he gained a forum in the now defunct magazine *Pensée*. Although Velikovsky was an extreme catastrophist, was his work unscientific? The most frequent criticisms of him are that his assumptions and reasoning were unsound, or simply that he was wrong. But many scientists reason by analogy, unfounded assumptions, and circumstantial evidence, and yet they are not barred from publishing in established journals. Nor has being wrong been an obstacle to publication: most articles published in decades past seem wrong to us now. Truth is hard to come by.

Of course, Velikovsky probably is wrong. It is impossible to imagine that one man could single-handedly refute most twentieth-century science without slipping up somewhere. And his supporters' claims that many of his 1950 predictions have been verified by subsequent research are simply false. The reason establishment scientists reject Velikovskyism is that they believe he was *so* wrong that they aren't interested in any research bearing on his hypotheses. Of course, that is not a very open-minded position to be taken by those who are supposedly searching objectively for truth.

The blame lies equally with Velikovsky, however. If it is not required that scientific research be correct or even tightly reasoned, it does matter that it be relevant and useful to the scientific process. It must be related to previous work or demonstrate past errors. Research that rejects the fundamental assumptions or paradigms of modern science (as Velikovsky's seems to) must show that the new assumptions provide a better explanation for all facts, not just a few. Yet Velikovsky refused to place his theories in a context that could broaden the perspectives of, or be useful to, other scientists. He never adequately defended the seeming incompatibility of his work with the laws of physics. If he had rejected those laws, it would have taken much more than his solitary efforts to devise an acceptable alternative paradigm. Against his "take it or leave it" attitude, it was understandable that scientists did not wish to devote precious space in their journals to publishing articles that had no bearing on their own work.

This does not mean that Velikovsky's challenge should be dismissed lightly. Do we really know that it is physically impossible for Venus to be in its present orbit, apparently stable for millions to billions of years, if it grazed the Earth only a few thousand years ago? When pressed, most specialists in orbital theory admit that theorems of orbital stability in systems as complex as the solar system have yet to be rigorously proven. And many of our other facts about the solar system are also built on assumptions not quite so fundamental as Newton-Einstein mechanics. We all ought to be more aware of our assumptions and more open to questioning them should contradictions appear. Or, at the very least, while most scientists pursue research guided by currently accepted paradigms, we should be tolerant of the few who march to different drummers.

For the near future, at least, Earth and planetary science will fashion its own mixed blend of uniformitarianism and catastrophism, rooted in the physical and chemical laws we think are fundamental. If and when our emerging model for the solar system's origin and evolution is found wanting, it may be overthrown as previous scientific paradigms have been (e.g., the recent acceptance of continental drift). It seems extremely unlikely to me that Velikovskyism will have been bolstered to the degree that it then will provide a superior model for the solar system, but who can say for sure?

4

Fragments from the Solar System's Birth

If an occasional asteroid were not a single body but consisted of several pieces . . . , we could never tell the difference.

—*N. T. Bobrovnikoff, 1929*

Our culture values greatness. Things small but beautiful never make the *Guinness Book of World Records.* Every touchdown leads inexorably to the Super Bowl. We extol superstars among us, one of whom stares from *Time*'s cover each new year as Man (occasionally Woman) of the Year. No wonder we all know of Jupiter (the largest planet), Venus (the nearest and brightest planet), and Saturn (the famous ringed planet). But who has heard of the planets Bamberga, Dembowska, Kleopatra, or Nysa? They also orbit the sun but are diminutive in size—just four among the countless asteroids that range in diameter from the breadth of Texas down to the mere pebbles and celestial sand grains we call meteoroids.

The French aviator and poet Antoine de Saint-Exupéry challenged our preoccupation with matters of greatness in his fable *The Little Prince*. His young hero lived on a small asteroid and cared for a single, beautiful flower. The prince traveled to Earth and en route visited some minor planets, numbers 325, 326, 327, 328, 329, and 330. It was with prescience that in 1943 Saint-Exupéry portrayed each as a small, unique world, with a single inhabitant of characteristic virtues and vices. The Little Prince met a benevolent king, a conceited man, a rationalizing drunkard, a greedy capitalist, a devoted lamplighter, and finally a narrow-minded scientific specialist. Befitting the variety of their occupants, some of the tiny worlds were bigger, others smaller; some spun rapidly, others more slowly; some had mountains, others did not; and they were of many different hues.

Only since 1970 have astronomers too begun to realize the variety of these little worlds and their significance for understanding our own. Although the discovery of the first (and the largest) asteroid in the year 1801 was greeted with enthusiasm, Ceres was disappointingly small for the planet that had been predicted to lie between Mars and Jupiter. As ever smaller asteroids were found, interest in them degenerated to a mere race among the more pedestrian observers to discover more than anybody else. Saint-Exupéry satirized these endeavors by remarking that the discovery report of "asteroid B-612" before an international conference was by an astronomer in "Turkish costume," so the report was not believed. More than one astronomer has called asteroids "the vermin of the skies" because they meander through familiar star patterns leaving unappealing trailed images on long-exposure photographs of galaxies and other (to some) more glorious celestial wonders.

The growing respectability of asteroid research is largely the result of the efforts of Tom Gehrels. Now a professor at the University of Arizona, Gehrels pays attention to details and, like the Little Prince, loves to travel to out-of-the-way places. It was on such a trip to America as a youth from the Netherlands that he wandered up Mount Palomar and was introduced to the astronomers by the groundskeeper, who had found him and put him up for the night. Nowadays he and his family may journey to Borneo or Sri Lanka, but his most contemplative and relaxing times are spent once again on Mount Palomar, where he periodically photographs the skies in search of faint and unusual asteroids.

During the 1970s, Tom Gehrels worked mainly with an instrument on the two Pioneer spacecraft that scouted the giant planets Jupiter and Saturn. As Principal Investigator of that experiment, he was responsible for some color photographs of Jupiter's famous Red Spot and the new Saturnian rings. Yet his work with Pioneer was but a passing fancy compared with his decades of research on minor planets, to which he has returned full-time. Starting in the mid-1950s, Gehrels and his mentor, Gerard Kuiper (the founder of modern planetary astronomy), patiently pursued a program of measuring the brightnesses and colors of asteroids. The modern realization that minor planets are a multitude of individual bodies, with different colors, different shapes, and different rates of spin, has grown from this program of photometry.

By recording the brightness of an asteroid with a photomultiplier—a supersensitive electric eye—attached to a telescope, astronomers can measure a "light curve." As an oblong body spins on its axis, it first seems brighter as it presents its broadest sunlit face to Earth, then fainter as we see it end on, then brighter again, and so on. Especially great brightness fluctuations reveal an asteroid's shape to be unusually oblong or irregular.

The Little Prince, who loved sunsets, could watch more than ten each 24 hours if he were on Icarus (asteroid 1566), but he would have to wait 85 hours between sunsets if he were on the larger, pinkish body called 182 Elsa. One tiny asteroid discovered in 1981 hardly spins at all: once in 148 hours! Only a mile across, it has been given the provisional designation 1981 QA until its orbit can be well determined. At that time it will be assigned a permanent number and its discoverer will be given the chance to pick a name for it. In the past, asteroids have been named for Roman gods, loved ones, cities, and even computers.

Tom Gehrels was ahead of his time. As late as 1970, by which time he had assembled a small group of students and Tucson, Arizona, high school teachers to assist his research, he remained the only astronomer in the world seriously studying the physical properties of asteroids. To be sure, he had gathered statistics on the spin rates and colors of dozens of asteroids, ranging in size from the small mountains hurtling past Earth to the much larger, more enigmatic "Trojan" asteroids orbiting the sun on each side of Jupiter in that planet's distant orbit. But what did all these data mean? What did they reveal about what asteroids were made of and how they were formed? Gehrels didn't really know, and at the time nobody else was very interested.

By 1970 a few graduate students at other universities had begun to take an interest in asteroids. The Nobel Prize–winning Swedish physicist Hannes Alfvén, who founded the esoteric subject of magnetohydrodynamics and was studying its possible ramifications for the origin of small bodies in the solar system, publicly rebuked the National Aeronautics and Space Administration for ignoring asteroids as potential targets of future spacecraft missions. Gehrels decided the time was ripe and, with Alfvén and several others, he organized a 3-day international symposium on the physical studies of asteroids. Held in Tucson in March 1971, the conference was attended by over 140 scientists, who presented 70 talks on various aspects of asteroid science. In the preface to the 700-page symposium proceedings, Gehrels wrote, "We are now on the threshold of a new era of asteroid studies. . . . Physical studies . . . have not been popular, at least not among astronomers. The lack of appreciation is coming to an end with the presently growing realization that asteroids, comets, and meteoritic matter are basic building blocks of the original solar nebula. Their exploration gives data for the study of the origin and history of the solar system. [This international conference was organized] to promote new and increased exploration [of asteroids]. . . ."

The conference was an exciting one for participants, who discovered the breadth of their mutual, latent interest in these neglected members of the solar system. Since then, Tom Gehrels's hopes have proven prophetic: his conference ushered in a period of frenzied research activity. Asteroids, once the purview of a single researcher, now account for about 20 percent of ground-based planetary astronomical research.

What caused the moon to melt early in its history? Why are there nickel-iron meteorites, hunks of solid metal fallen from the skies and used in centuries past both as anvils and as objects of devotion? How were the precious pre-solar-system remnants of stellar explosions, which some researchers have detected in meteorites, protected from the heat and collisions that melted, vaporized, or ground down to nothingness so much material once part of the early solar system? How did the vast whirlpool of dust and gas that surrounded the sun at its birth manage to agglomerate and coalesce into just a few dozen large planets and satellites? Only in the decade since Gehrels's symposium have astronomers fully appreciated that precious clues concerning such diverse questions reside in the asteroids—those neglected, rocky fragments beyond the orbit of Mars.

Consider the sizes of the minor planets. The largest is Ceres, with a diameter of about 1,000 kilometers.* Two others exceed 500 kilometers. But 30 or 40 exceed 200 kilometers, the length of Massachusetts excluding Cape Cod, and roughly 3,000 exceed 20 kilometers (the size of Lake Tahoe). Literally trillions of uncharted boulders the size of a basketball or larger exist. For each asteroid there are about 10 others one-third its size. It is difficult to visualize the ever-increasing numbers of smaller bodies. After all, most familiar objects in our world are of similar sizes. Few apples in a basket, or pigeons in a park, differ in size by as much as a factor of 2. Yet one familiar process does produce an asteroid-like size distribution. When your children smash a window or priceless vase, count the broken fragments. Or smash a brick with a sledgehammer. The more you repeatedly grind something to

---

* A kilometer is a bit more than half a mile, a centimeter a bit less than half an inch. I use metric and English units interchangeably in this book. We must soon get used to metric units, but it is hard to switch instantaneously.

pieces, the more nearly the fragments match the asteroidal size distribution.

So asteroids would seem to be fragments. Indeed it had been supposed since the first ones were discovered that they might be pieces of an exploded planet. The Canadian expert on celestial mechanics Michael Ovenden has suggested that the asteroids are the remains of a giant Saturn-sized planet he calls Aztec, which mysteriously self-destructed about the time of the emergence of hominids on Earth, 10 or 20 million years ago. It boggles the mind to imagine what kind of catastrophe could blast apart such a huge planet, so tightly bound together by its own gravity. But there are other incredible aspects to Ovenden's scenario. Since the total of all asteroids today makes up only 1/100,000 of the proposed mass of Aztec, 99.999 percent of it must have been flung about the solar system and eventually ejected into interstellar space. Yet lunar lava flows, which have been passively recording impacts for more than 3½ billion years, have been cratered only infrequently, and collected lunar soils show no chemical trace of Aztec. How could the moon have remained unscathed? Why do the remaining asteroids orbit the sun in the same direction in a relatively orderly ring, in roughly circular orbits, rather than scatter about in highly inclined and elliptical paths intersecting at the site of the supposed explosion?

Actually the fragmental nature of asteroids is much more simply understood as resulting from a kind of celestial demolition derby. One can calculate, simply by knowing how many asteroids there are and how fast they are moving, how often bodies of different sizes will collide with each other. If such collisions smash asteroids to pieces, then they must be slowly grinding themselves down to dust. Imagine you were to enter an arena to watch a dozen driverless jalopies careening about. Wouldn't you be surprised if the automobiles were all functioning and undented? Not if the event had started only seconds before you arrived. But after half an hour of this madness, most vehicles would certainly be out for repairs. Similarly, there are so many asteroids flying about in space at relative speeds of 5 kilometers per second that nearly all of them must have suffered catastrophic collisions with each other over the eons. Perhaps there were many more asteroids when the solar system formed, but most have smashed each other to bits because of the crowding, leaving only the asteroids we see today.

Although it is easy to figure how *often* asteroids collide, what really *happens* when they do? What does it take to smash an asteroid apart? Difficult as it is to conceive of such devastating collisions, it is quite impossible to experiment with processes vastly more powerful than nuclear explosions. However, scientists can study how rocks fragment and the physical laws involved and, by extrapolating, they can begin to picture what happens. Of course, it depends on the size and impact velocity of the smaller asteroid: the larger and faster this projectile, the more catastrophic the result. Also important is the size of the target body and, for a small target, what it is made of. Oblique impacts on a rapidly spinning target may have especially remarkable results.

Suppose an asteroid the size of Rhode Island ploughed at 5 kilometers per second into one the size of West Virginia. Of course, there would be a tremendous, blinding explosion. "Rhode Island" would be mostly melted and vaporized into oblivion. The energy of impact also would be more than enough to smash "West Virginia" into pieces, and propel the fragments up and away at 200 miles per hour. You might think that would be the end of "West Virginia," but not so! Escape velocity is still greater than 200 miles per hour, so most of the fragments would circle around, fall back together, and re-form the asteroid into a huge, gravitationally bound pile of boulders and rubble. Had the projectile been as large as Connecticut, "West Virginia" would have been permanently dismembered as well as shattered; its fragments would have gone into separate, but similar, orbits around the sun. Many such asteroid "families" exist in the belt, testimony to earlier catastrophic disruptions.

Probably all but the very largest asteroids have had shattering collisions during their lifetimes. But few of the many hundreds of asteroids larger than 50 kilometers in diameter have yet been catastrophically disrupted. So, battered and bruised as they are, the larger asteroids are mostly "original" bodies. Only those smaller than about 25 kilometers in diameter are likely to be pieces of larger disrupted parents.

Few collisions are exactly head-on. A series of oblique impacts on a large asteroid tends to spin it up faster and faster, with strange results. When it is rotating every 5 hours or so, it develops an equatorial bulge. If it is hit again on its receding side by an impact hard enough to pulverize and jumble the body but not destroy it, the fragments reaccumulate into an elongated ellipsoid, nearly three times as long as it is across. Now it is spinning so fast, once every 4 hours, that surface rocks near its ends are in danger of being flung into space, just as centrifugal force on a fast merry-go-round nearly overcomes a child's grip.

If such an ellipsoidal asteroid were struck on its receding side yet again, the fragments might have too little energy to reach escape velocity but too much angular momentum to be able to reaccumulate into one whirling rubble pile. My colleague Stuart Weidenschilling of the Planetary Science Institute believes a double asteroid would result, or perhaps one with a retinue of fragmentary satellites! This is not the kind of world encountered by the Little Prince in his travels, nor is it a configuration astronomers thought could exist until a few years ago (despite the early insight quoted at the start of this chapter). A few years ago observers started to measure asteroid diameters by timing how long a bright star disappears when an asteroid passes between the Earth and the star. Unexpectedly, they saw stars disappear more than once! Their reports were greeted with skepticism, but more recently sophisticated "pictures" have been taken of at least two asteroidal satellites, resulting in a minor revolution in asteroid science.

It is not known how many asteroids might be double or have satellite systems (or even rings!). It appears possible for an occasional asteroid to undergo a series of oblique impacts sufficient to spin it up to a state of duplicity before it is catastrophically disrupted. A small percentage of asteroids may become double this way. If many asteroids are found to be double or have satellites, then we must figure out how asteroids might have originally formed double.

As I said earlier, the outcome of a collision involving an asteroid too small to have much gravity depends on what it is made of. Small asteroids could be fragments of larger ones *if* asteroids are made of ordinary rock. Were they as strong as steel, however, few collisions could shatter them. Apparently most asteroids are, in fact, rocky. We have come to realize that we even have pieces of asteroidal rocks right here on Earth—in our museum collections of meteorites, the stones that have fallen from the skies.

How can we tell what a distant asteroid is made of? Only with great difficulty, is the answer. The technique is called "remote sensing," which is the art of measuring light reflected from, or radiation emitted by, a surface in order to learn what it is made of and what it is like. Some Earth-scientists use satellite remote-sensing data to prospect for ores, detect forest blights, monitor lake pollution, or even locate illegal crops. They can test their interpretations by traveling to see what is really there. From such "ground truth," they develop a feeling for what they can—and cannot—interpret reliably from satellite data. Astronomers do remote sensing too, but there's a difference. Except for the Apollo moon-landing sites, they lack access to the ground truth on stars or planets. The calibrations by Earth-scientists of plants, soils, and pollution are of little help to astronomers dealing with strange worlds having un-Earthly environments. Nevertheless, as one planetary astronomer has said, "Remote sensing is the next best thing to knowing what really is there." The approach has been applied to asteroids with some success.

We see planets and asteroids by the reflected sunlight that illuminates them, but the sun's rays are modified while interacting with their surfaces. On a microscopic scale, the rays bounce around among the mineral grains on the rock surfaces and are transmitted through some of them before traveling back to our eyes. Some rocks absorb most of the blue, violet, and ultraviolet rays and transmit easily only the orange and red rays; thus they appear reddish. Others strongly absorb the rays toward both the violet and red ends of the spectrum, yielding a greenish hue. Anyone who has hiked into the Grand Canyon, or just strolled along a rocky stream, has noticed the variety of colors of rocks on Earth. Such colors ultimately result from the atomic structure of the mineral crystals of which rocks are made. The outer electrons, particularly of iron, cobalt, and similar atoms, are unusually sensitive to light of particular wavelengths in the red and invisible infrared; the exact hue of absorbed light depends on the crystal structure of the particular mineral. Scientists can infer the minerals that compose rocks on remote planets just by measuring the wavelengths of the more strongly absorbed reflected sunlight. Actually, to measure these spectra is not an easy task. Instruments must be built that are much more sensitive to fine gradations in colors than is the human eye. Such spectrophotometers have been used on many of the great telescopes atop mountains in Hawaii, South America, and the Southwest, including the 200-inch Hale telescope on Mount Palomar.

The power of this research technique in planetary astronomy was first demonstrated by Thomas McCord, then of M.I.T., but now working in Hawaii. He proved, several months before U.S. astronauts returned the first rocks to Earth, that the moon's surface is made, at least in part, of the very same minerals that compose rocks on Earth! Of course, the major minerals in lunar rocks occur in different proportions than in the Earth's rocks, revealing clues about the nature of the moon (see Chapter 8).

Spectrophotometry of asteroids has shown that some contain minerals common in certain terrestrial rocks, but others are quite exotic. Consider asteroid 324 Bamberga. Perhaps the Little Prince began his travels with Number 325 because he couldn't even see Bamberga, which is dim and seems unexceptional. But in 1970, when then-Caltech graduate student Dennis Matson measured the heat being radiated toward Earth by Bamberga, he found it was one of the brightest sources of thermal radiation in the sky, which meant that it was a warm and unexpectedly large body. For Bamberga to be large and warm but dim to the eye means that it must be exceptionally black. In fact, it absorbs 98 percent of the sun's rays of all colors, from ultraviolet straight through infrared. Bamberga is blacker than coal dust!

What can a 250-kilometer-diameter planet be made of that is that black? Clearly it must contain chemical elements that cosmochemists believe are reasonably abundant in the universe. The only common element that frequently forms black compounds is carbon, which is fourth in cosmic abundance, after hydrogen, helium, and oxygen. Yet even on Earth black carbon compounds are quite rare and usually are associated with decomposed life; from them we make "lead" pencils, black ink, carbon paper, and so on. But there is one kind of black, carbonaceous rock that falls from the sky: the carbonaceous chondritic meteorites.* These meteorites are rarely found because they are so fragile that most of them totally disintegrate while decelerating through the Earth's atmosphere. Yet they are believed to be the most common kind of rocks in space. Laboratory measurements of the light reflected from powdered carbonaceous meteorites show that it closely matches Bamberga's spectrum. So it seems likely that Bamberga and the many other similarly black asteroids are giant carbonaceous chondrites and that the carbonaceous meteorites in our museums are chips from some of these asteroids—extraterrestrial samples even more unearthly than moon rocks!

Rocks in the Earth's crust are rich in silicon and aluminum. Spectrophotometry reveals, however, that such rocks are uncommon on asteroids. A couple of highly reflective asteroids seem to be made of whitish rocks that are similar to the unusual aubrites (a type of achondrite). Vesta, the third largest asteroid, is made of rock that resembles the volcanic lavas that spread across the surfaces of the moon and Earth; in this respect Vesta is unique

* *Chondrites* are overwhelmingly the most common kind of stony meteorite. They are named for *chondrules*—tiny spherules about a millimeter across—which comprise much of the volume of so-called *ordinary chondrites*. Ironically, *carbonaceous chondrites* generally lack chondrules, whose origin is in any case a mystery. Whether they contain chondrules or not, chondrites have proportions of nonvolatile chemicals similar to chemical abundances in the sun and other stars. Thus they are thought to be representative of the material from which the planets were made. Rocks like those found on the Earth and the moon (and like the *achondritic* meteorites) exhibit very nonsolar abundances, reflecting planetary processes of chemical segregation. (I am sorry about the jargon, but simpler words for meteorite types have not been invented.)

among asteroids, but as I will describe shortly, there may have been other Vesta-like bodies in earlier epochs.

The second most common type of asteroid—carbonaceous ones are by far the most common—are moderately reflective and have the iron- and magnesium-rich silicate minerals called pyroxene and olivine. In that respect they are just like the ordinary chondrites, the most common meteorites to survive a fiery fall through our atmosphere. These so-called S-type asteroids also have a reddish tinge, probably due to a pure nickel-iron metal alloy, unlike anything found naturally on Earth (except of course in the iron meteorites).

The third most common type of asteroid also exhibits the reddish tinged spectrum, but shows no evidence of other minerals. Perhaps they are made of solid metal, in which case there might be tens of millions of billions of tons of nickel-iron alloy and associated precious metals in the asteroid belt. The economic potential of this storehouse of metal, in the event mankind conquers and industrializes space, is staggering. But the implications of these interpretations for the early history of the solar system are also fascinating and important.

Back in the early 1970s, asteroid observers were busy characterizing the larger asteroids as being of carbonaceous composition (called C-types), silicate plus metal (the S-types), or less common types. At the same time, asteroid sizes were being determined reliably for the first time. One afternoon in August 1974, I graphed the sizes of the S-type asteroids separately from the C-types and I noticed that most S-types were between 100 and 200 kilometers in size, with a curious lack of many smaller or larger ones. The C-types seemed to follow the normal fragmental size distribution that holds for a broken vase.

It occurred to me that differences in composition might account for the size differences. At the time, Thomas McCord believed that the reddish tinge of S-types meant their surfaces contained at least as much metal as rocks. In fact, he and Michael Gaffey suggested they might be like the so-called stony-iron meteorites. These consist of rocks that were invaded by molten metal, which subsequently cooled into a solid metallic network. Such meteorites probably formed at the boundary between the metallic cores and rocky mantles of smallish planets. Imagine that there were once parent bodies, 300 to 500 kilometers in size, composed of ordinary chondrites. The meteorites, which contain metal grains in addition to pyroxene and olivine, are thought to be among the materials that originally condensed from the solar nebula—the cloud of gas and dust from which the solar system formed. If somehow the parent bodies had been heated and melted, the dense grains of metal would have sunk to their centers, forming molten nickel-iron cores. Geochemists calculate that lighter magmas would have floated and lava might have flooded out onto the surfaces of such bodies, just as on Vesta. Later, the bodies would have cooled and solidified throughout, including their metallic cores 100 to 200 kilometers in diameter.

On that broiling August day in Arizona, my thoughts turned to memories of my boyhood winters in Buffalo, New York, for an analogy that might explain the asteroid sizes. Two brothers who lived on the block had tendencies toward delinquency. They weren't satisfied, as the rest of us were, to throw fluffy snowballs at passing cars. They froze their snowballs into iceballs. Occasionally they made snowballs with rocks in the centers. I well remember my playmates being chased through the neighborhood by irate drivers of dented cars. The asteroid analogy, however, concerns the fate of the snow and the central rocks. The snow, of course, splattered everywhere when the missiles struck a car, but the rocks remained intact. It seemed to me the same might be true for asteroids, whose solid metallic cores were much stronger than their rocky outer layers. Over eons in the celestial demolition derby, the odds would be against survival of the rocky outer layers of the once-melted Vesta-like asteroids, but their strong metallic cores would have remained whole: split apart at their core/mantle interfaces, they would appear to be stony-iron asteroids of 100 to 200 kilometer sizes.

I published an article on my idea. Then it began to fall apart. As more asteroids were classified and measured, the differences in size distributions of S- and C-type asteroids vanished. There was no predominance of 100- to 200- kilometer S-types to explain, after all. Then my associates and I began to fully appreciate how gravity fields keep larger asteroids from disrupting. It now seems very unlikely that rocky mantles more than 100 kilometers thick could have been stripped from imbedded cores.

In 1981 a graduate student at the University of Arizona, Michael Feierberg, studied new infrared spectra of S-type asteroids and concluded that they are not like stony-iron meteorites, after all. He thinks that their reddish tinge is consistent with both the 10 to 20 percent metal content of ordinary chondrites and with the greater than 50 percent content of stony-iron meteorites. But all ordinary chondrites have mixtures of the two silicates pyroxene and olivine, while stony-irons are either metal mixed with pyroxene or metal mixed with olivine, but *never* metal mixed with both. Although it is difficult to determine the metal content of an asteroid from its spectrum, Feierberg found that spectra of most S-type asteroids show absorptions due to both olivine and pyroxene, so they probably are ordinary chondrites.

The nugget of my snowball idea may yet be salvaged, however. Although Vesta-sized bodies cannot be stripped to 150-kilometer cores, my associate Richard Greenberg and I now think that stony-iron meteorites may be derived from denuded 30-kilometer-diameter cores of parent bodies originally less than 100 kilometers in size. If a few percent of the asteroids of all sizes had melted and formed cores, most of the smaller ones *would* have had their rocky mantles stripped away. Subsequent impacts on the resulting small core bodies might have chipped off metallic and stony-iron fragments, thus producing some of the meteorites that have fallen on Earth. Our hypothesis accounts for many characteristics of meteorites. Whether smaller metal-rich asteroids really exist with the sizes and compositions we predict must await future observations.

During the decade since Tom Gehrels's meeting, astronomers have learned what the asteroids are like: how big they are, what they are made of, and so on. Indeed, we have learned that most meteorites almost certainly are derived from the asteroids—a venerable hypothesis that nevertheless was doubted by most theoreticians only a decade ago. We also have learned how the asteroids behave today, especially about the occasional collisions that, multiplied over eons, have pulverized them. We even have ideas about what the asteroids were like several billion years ago. But this knowledge only sets the stage for addressing more fundamental questions: Why is there an asteroid belt, rather than a planet, between Mars and Jupiter? Why were some asteroids heated to the melting point while others seem to be composed of unmodified material from the solar system's birth? At last we can make some informed speculations, but final answers await research in years to come.

Let me portray a scenario for the origin of the asteroids and the larger planets. We are confident of parts of this picture, but many mysteries remain. The sun was born from a giant cloud of gas and dust left over from some exploding stars in the Milky Way galaxy. This cloud, called the solar nebula, contracted and the internal gas pressure increased; hence the temperature rose. By the time a protosun appeared, the swirling nebula had become a revolving disk of gas with temperatures approaching a couple of thousand degrees in the zone of the yet unborn inner planets. As it cooled, the gas began to condense into grains of minerals that can exist at the highest temperatures. Farther from the sun, the nebula cooled so that a wider variety of minerals condensed where the Earth and Mars were soon to form. Ordinary chondrites condensed near the inner edge of the asteroid belt, where S-types are still common. Evidently carbonaceous chondritic matter, which condenses completely only at temperatures below a few hundred degrees, formed in the rest of the asteroidal zone. Farther outward it was still colder so that ices formed, leaving only the lightest and most volatile substances in the gaseous state.

The grains grew from the condensing gas and, just as in a snowstorm, they began to settle and fall to the central plane of the nebular disk. Remaining gases were dispersed by the youthful sun; meanwhile instabilities in the central disk and the mutual interactions of the crowded grains caused them to gather together into small spheres, several kilometers in diameter, called planetesimals. Planetesimals near the sun, where Mercury was to form, consisted of only a tiny fraction of the original nebular gases—materials, including iron, that condense at the hottest temperatures. But in the outer solar system, where even the remaining gas had not been blown away by the distant sun, space was crowded with icy planetesimals that bumped into each other and grew in size. One of them, a proto-Jupiter, grew to the size of Ceres; its gravitational field began to attract gases and other planetesimals. In a cosmologically brief time the mighty planet Jupiter had formed, perhaps leaving behind some large, partly formed, stray planetesimals in nearby orbits.

Meanwhile, other planets slowly accreted from swarms of planetesimals elsewhere in the solar system. For a planet to grow, it was necessary for the planetesimals to move with respect to each other, in similar but not identical paths; otherwise they would never bump together. They could not be speeding too fast or they would fragment rather than merge together. So the asteroidal planetesimals must originally have been moving in similar, circular orbits, inclined only slightly to the flat central plane of the planetesimal system, just as one might expect for small bodies born enveloped in a rotating gaseous disk. Yet today asteroids flash by each other at speeds of many kilometers per second and smash each other when they collide. What stirred up their once orderly orbits, and was this mysterious occurrence the reason why the asteroids never formed a planet? Nobody knows for sure, although we are beginning to narrow down the admissible hypotheses.

Most scientists think Jupiter must somehow have been responsible for the failure of the asteroids to accrete, as well as for the small size of the next-closest planet, Mars. Jupiter is so massive that its gravitational field can be felt at great distances. At certain distances from Jupiter, orbiting bodies pass by just often enough to receive periodic gravitational tugs from Jupiter that build into large changes in their paths. Indeed, the regions in the asteroid belt that would have been inhabited by bodies orbiting with periods commensurate with Jupiter's mighty pulls are altogether empty. But in most parts of the asteroid belt, Jupiter's tugs occur at the wrong times and are not amplified into large excursions, so Jupiter's effects are calculated to be minimal. Perhaps the narrow "resonance zones" of Jupiter's effects once migrated throughout the asteroid belt while the sun was divesting itself of its nebular disk. Then Jupiter might have stirred up all of the asteroids.

Other theories suggest that the asteroidal planet was stillborn because of a cosmic shooting gallery created by Jupiter. Icy planetesimals left over in the outer solar system after Jupiter had formed chanced to come near the giant planet; they were flung off in divergent paths, just as Jupiter tossed Pioneer 10 into interstellar space after its reconnaissance flight several years ago. Such speeding planetesimal-bullets entering the inner solar system might have been especially effective in smashing some of the nearby growing asteroids and Mars-zone planetesimals, thereby inhibiting growth.

Such collisions alone cannot account for the asteroids' divergent orbits. Imagine being given an arsenal of guns and being told to move a large glass sphere down the length of a football field by firing at it. If you fired at it with a BB gun, it would move hardly at all; struck by a rifle bullet, the sphere would certainly be smashed to bits long before you got it to the 50-yard line. Similarly, asteroids are fairly fragile and cannot be shifted intact by collisions. The stirring of the asteroids' orbits must have resulted from strong gravitational forces, which operate gently over longer durations than violent impulses. If one of the planetesimals scattered by Jupiter were as large as Mars or the Earth, its gravity field might have done the trick. It would have followed an elongated orbit crossing the asteroid zone for 10 million years before again encountering Jupiter and suffering Pioneer 10's fate. That is long enough for it to have passed close to most of the asteroids and to have accelerated them. This disruption of their orderly orbits eliminated the possibility of any asteroidal planet.

We are fortunate that Aztec never formed, for we are thus left with samples of planetesimals like those that originally formed the Earth, Mars, and Jupiter. Asteroids are somewhat battered, so the clues are a bit difficult to decipher. But all other evidence of this critical stage of planetary evolution has been totally lost forever: other planetesimals were buried deep in the growing planets, melted, and transformed beyond recognition by the great pressures, geological forces, and chemical reactions endemic to large planets. The astronauts who journeyed to the moon hoped to bring back rocks from the earliest years of the solar system. But even our relatively small moon turned out to have been largely melted. Whatever early lunar rocks might have avoided melting are buried deep beneath kilometers of rubble on the lunar surface, since virtually all the rocks excavated by cratering impacts fall back onto the moon, rather than escape into interplanetary space as they do from a small, low-gravity asteroid. So the asteroids are the true Rosetta stones of the inner solar system, just as comets may be for the regions beyond Jupiter. Many asteroids probably contain rocks formed at the very beginning, when the solar nebula was beginning to condense. Most meteorites, the asteroidal fragments, date from roughly 4.6 billion years ago, making them more than 600 million years older than many lunar rocks.

Most asteroids have been unmodified by thermal or chemical processes since the early condensation and accretion, but there are important exceptions. Remember that a few percent of asteroids were heated so much that they melted and the minerals segregated into metallic cores surrounded by layers of lighter rocks. Why did they melt? And if some melted, why didn't the others? The answers may have wider implications for all planets. Asteroids are not easily melted since they are too small to generate high pressures in their interiors and they radiate away radioactive heat almost as fast as it is produced. So if something melted asteroids, it probably would have had an even more important effect on heating the young moon, Earth, and other planets, if they had formed by that time.

The sun, even if it were once much brighter than today, can warm only the uppermost layers of an asteroid. The radioactive decay of uranium, thorium, and potassium produces heat which, if trapped deep inside a planet, can eventually heat it to melting. This is the reason for the great internal warmth of the Earth today. But just as a thin-walled, poorly insulated house quickly loses heat during a cold night, tiny asteroids cannot be heated much by such decay of long-lived radioactive elements, despite their partially insulating fragmented state.

Some researchers hunting for the ancient furnace have turned to giant electrical and magnetic fields that may once have been carried past the asteroids by an interplanetary windstorm originating on the sun. Even today there is a stream of subatomic particles flowing away from the sun, and astrophysicists think this solar wind may have been much greater in our star's early history. In that case, calculations show that asteroid-sized bodies could have been heated by the large currents induced by the passing magnetic fields, in a manner analogous to the way the elements on an electric stove are heated by flowing currents. A novel trait of solar-wind heating is that asteroids of particular sizes and compositions may have been heated while others were little affected.

Another promising source of early planetary heating is the decay of radioactive elements having very short half-lives and which are therefore now extinct. For instance, one extinct form (isotope) of aluminum that contains 13 neutrons instead of the usual 14 tends to change spontaneously into a stable form of magnesium on a time scale of about a million years. Heat is a by-product of the decay. The unstable form of aluminum may well have been created by the stupendous stellar explosions, or supernovae, that may have helped create the cloud from which the solar system formed. If the solar system formed quickly enough—within a few million years of a supernova—the rapid radioactive decay of a sufficient quantity of the unstable aluminum might have provided a burst of heat for the newly formed planets in which it resided. Even small asteroids are large enough to retain heat for a few million years, if not for billions of years, so they also would have warmed and perhaps melted as a result. We can check to see if that happened by looking for unusual concentrations of the stable magnesium decay product in meteorites and lunar rocks, especially in rocks rich in aluminum but lacking the normal type of magnesium.

Research several years ago seemed to rule out such an early heat pulse, but more recently several research groups have reported anomalies in the proportions of magnesium isotopes in some meteorites that seem to be the long-sought clue to the early thermal history of the planets. Therefore, it is especially important to understand why some asteroids were heated to melting temperatures of over 1,000 degrees Kelvin while others never exceeded their condensation temperatures of a few hundred degrees. Hanging in the balance are some fundamental questions: How long did it take the solar system, and the various planets in it, to form from the debris of a stellar explosion? And how, during the first 10 percent of the solar system's history, did asteroidlike planetesimals get together to form the planets—large worlds with the heterogeneous environments necessary for the evolution of life?

1. Almost everywhere on Earth rainfall and running water shape the land. But here, in the Iqa Valley of Peru, virtually no rain falls and the landforms are shaped mainly by blowing winds and sand. These elongated ridges, several kilometers long, are called yardangs; their aerodynamic shapes resemble the bottom of a boat. While on Earth they are confined to the driest deserts, yardang-like features are common on Mars. *Courtesy NASA*

2. The Elegante crater lies in the Pinacate lava fields, south of the Arizona-Mexico border near the Gulf of California. Unlike the smaller cinder cones, or volcanic mountains, visible at the top, Elegante bears a superficial resemblance to meteor-impact craters. But it, and many similar craters in the Pinacates, is known to be of volcanic origin, so it is easy to understand why debate raged so long about whether the lunar craters are of impact or volcanic origin. *Courtesy S. Larson*

3. The Mariner 10 spacecraft. The cameras that took pictures of Mercury and Venus peer down from the top of the device. The "wings" are solar panels, each nearly 3 meters long. *Courtesy NASA*

4. Discovery Scarp is the prominent feature that runs north to south in the center of this Mariner 10 photograph of Mercury. It was named after the sailing ship used by Captain Cook on his 1776–80 explorations of the Pacific, Canada, Alaska, and Siberia. It is one of the numerous great wrinkles on the surface of Mercury that testify to the contractions of the planet as it cooled. *Courtesy NASA*

5. The left half of this mosaic of Mariner 10 photographs is dominated by the huge Caloris Basin, which is over 1,300 kilometers in diameter and is the largest basin on the side of Mercury observed during the mission. The basin's floor is covered with ridges and canyons. The lightly cratered plains to the right may have formed at the same time as the basin. *Courtesy NASA*

6. This odd-shaped panorama of rocks on the broiling surface of Venus is one of only two pictures ever taken from the surface of our neighboring planet. It was taken by the Russian lander Venera 9 before it succumbed to the heat. *Courtesy S. Nozette*

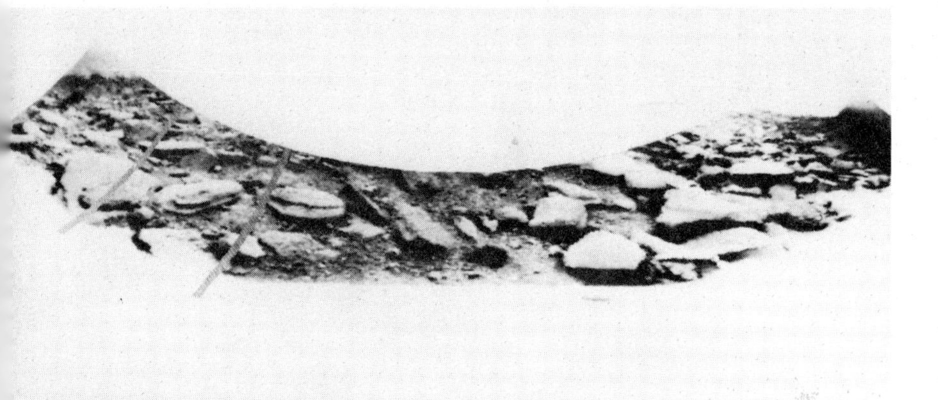

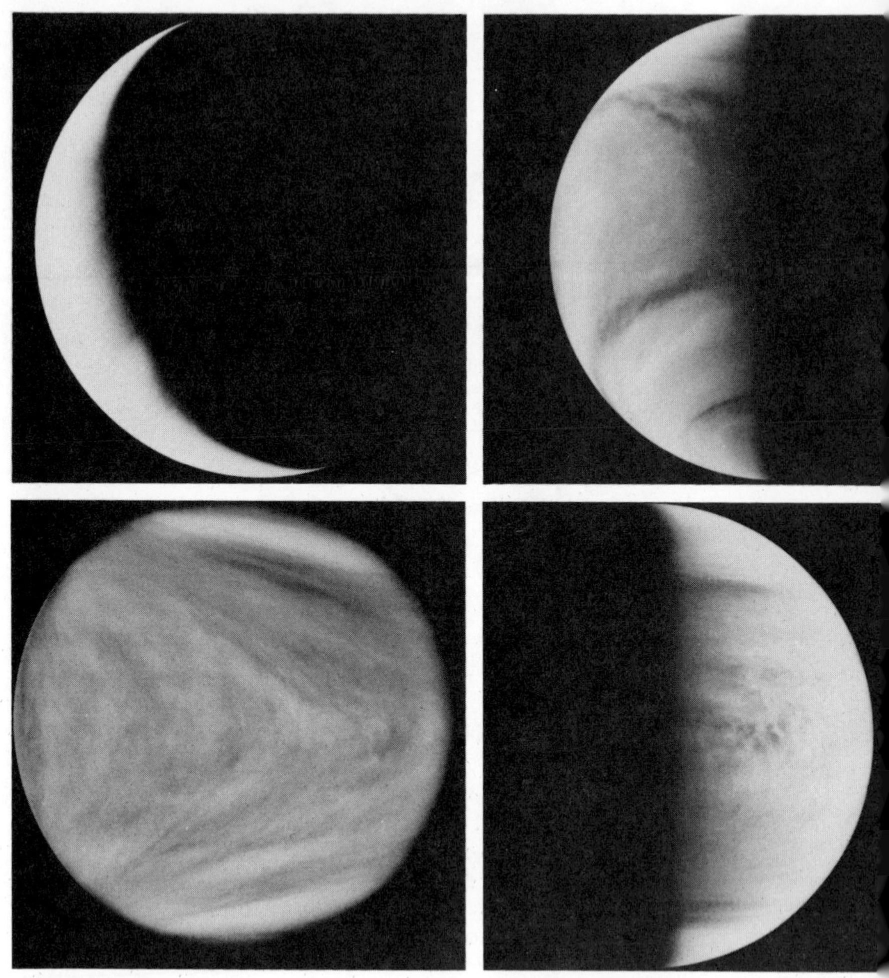

7. The phases of Venus are shown in these four pictures snapped by the Pioneer Venus orbiter December 1978–April 1979. They show the patterns of ultraviolet-colored clouds high in the upper atmosphere of Venus. *Courtesy NASA*

8. This oblique view of the Eratosthenes crater was snapped from the moon-orbiting Command Service Module of the Apollo 12 lunar-landing mission. The crater is surrounded by clusters of smaller, secondary craters formed by the impact of material ejected during the formation of Eratosthenes itself. *Courtesy NASA*

9. A chain of craters photographed from the orbiting Apollo 14 Command Module. *Courtesy NASA*

10. An unfamiliar view of the moon. This far-side photograph was taken toward the end of Apollo 16's lunar visit. *Courtesy NASA*

11. The cloud-bedecked planet Earth rises over the moon's horizon, as viewed from Apollo 17's lunar orbit, shortly before it returned to Earth in December 1972. That was the last manned expedition to the moon. *Courtesy NASA*

12. Scientist-astronaut Harrison H. Schmitt examines a large split boulder lying near the Taurus-Littrow landing site of the Apollo 17 Lunar Module. Schmitt later became a U.S. senator from New Mexico and has advocated a renewal of the space program. *Courtesy NASA*

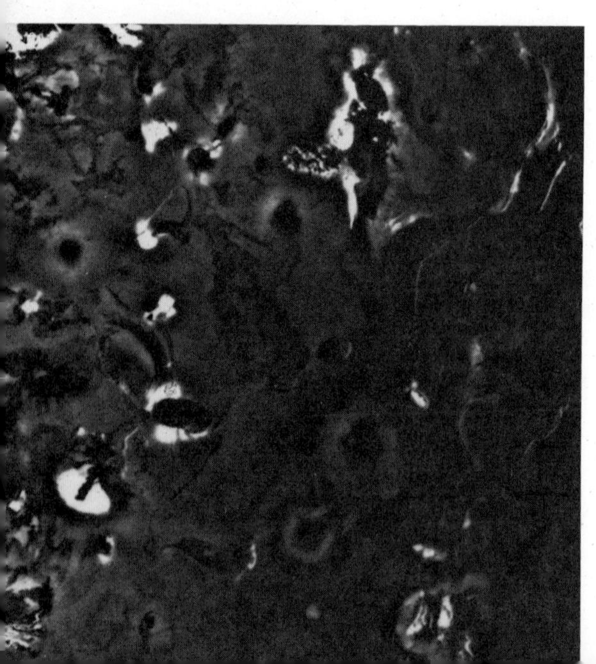

13. Close-up of the remarkable, sulfurous surface of Io, taken by the cameras of Voyager 1 during its historic encounter with the Jupiter system in March 1979. The small, dark, spidery feature to the left probably marks sulfur flows that have oozed from the small volcanic crater within it. *Courtesy NASA*

14. Perhaps the strangest and certainly the most geologically active world in the solar system is Io. In the center of this global view is a fuzzy, bright, circular ring. That is the plume of the sulfur-volcano Prometheus, which was actively fountaining when this picture was taken. *Courtesy NASA*

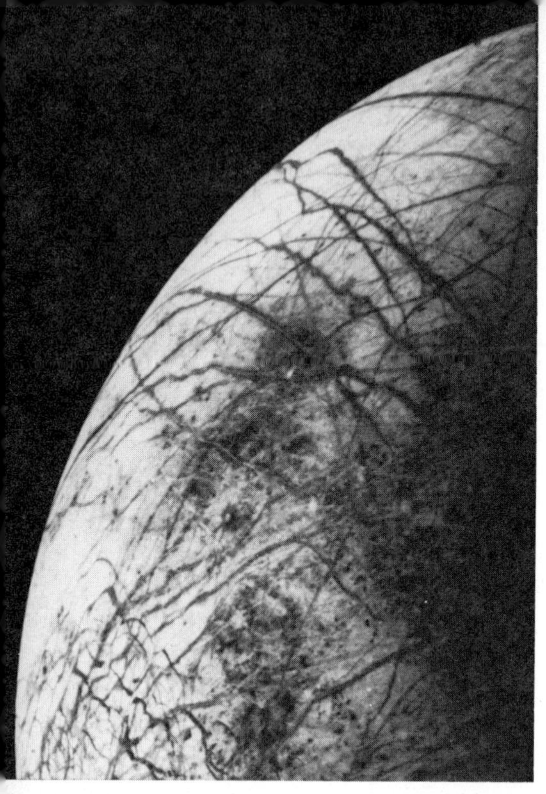

15. Long dark bands criss-cross the surface of Europa in this Voyager 2 portrait of Jupiter's second moon. Note the thin white lines within the larger bands. They are a clue to the origin of the stripes, perhaps by a process of explosive, aqueous volcanism.
Courtesy NASA

16. Europa is the smoothest world yet photographed in the solar system. Even near the "terminator" (day/night boundary), where shadows lengthen and the setting sun should highlight even lowly hills, Europa's surface seems nearly smooth, as seen in this Voyager 2 rendition. Scientists are baffled about the origin of the narrow, scalloped ridges that cross this part of Europa.
Courtesy NASA

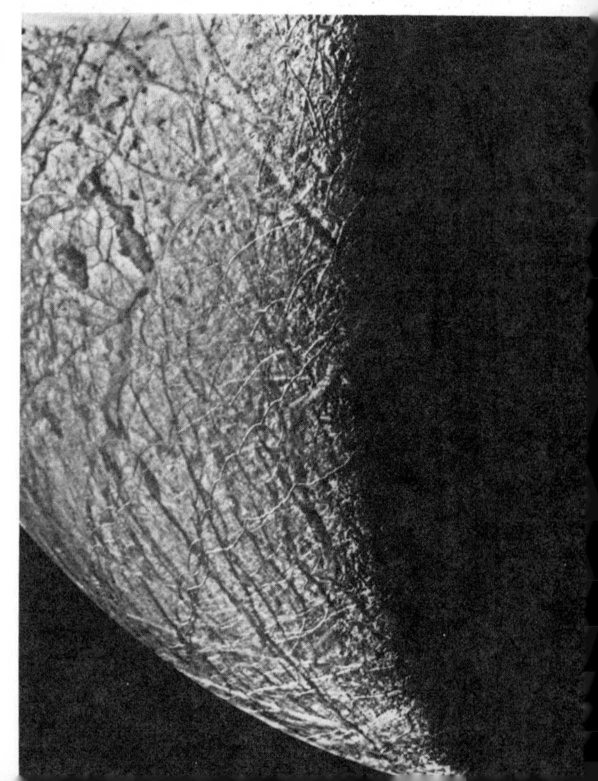

17. A global portrait of Jupiter's third Galilean satellite, Ganymede. The dark, circular "continent" is Galileo Regio, which is quite densely cratered on closer inspection. Its northern portions are brighter, due to polar deposits of some kind. The brighter bands on Ganymede are the terrains that are seen to be extensively grooved in higher-resolution pictures. *Courtesy NASA*

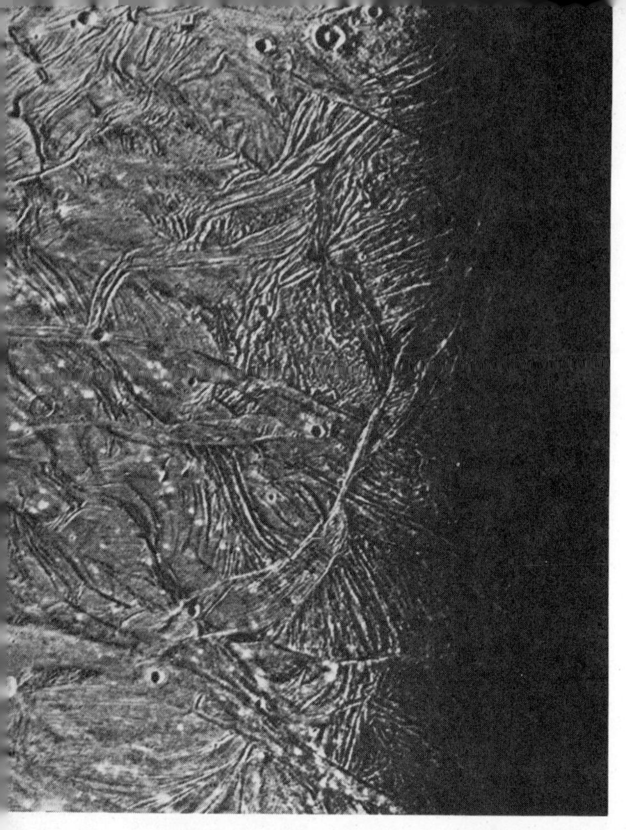

18. Overlapping patterns of grooved terrain on Ganymede highlight geological processes that may be uniquely characteristic of icy worlds. *Courtesy NASA*

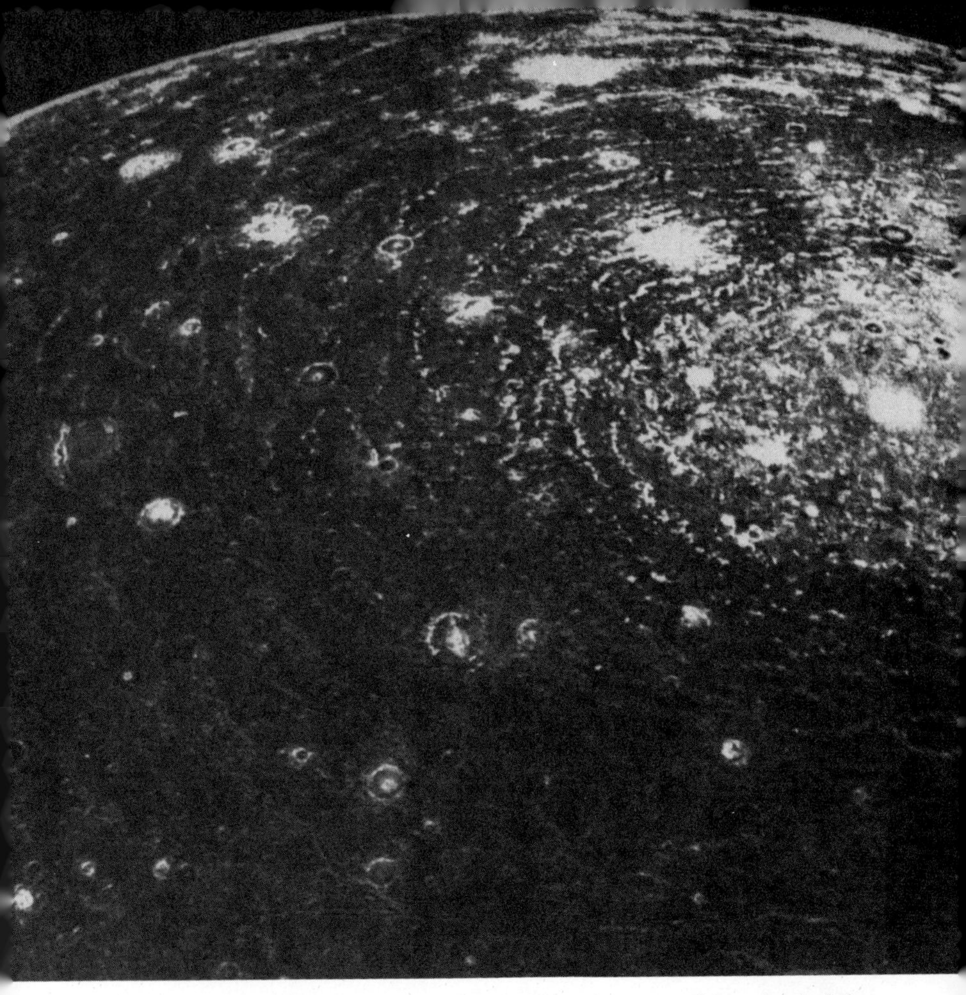

20. This entire Voyager 1 frame of Callisto is covered by the bright, concentric rings and escarpments that surround the old, giant impact feature named Valhalla. Although Callisto is generally a very dark-colored world, recent impacts and rock-slides have exposed fresh, white ice, which highlights some of the topographic features. *Courtesy NASA*

19. *(opposite)* The outermost of Jupiter's four Galilean moons, Callisto, is perhaps the most densely cratered planet-sized body in the solar system. But, as this Voyager 2 close-up shows, Callisto is remarkably lacking in its share of larger craters, by comparison with the moon, Mercury, or Mars. Scientists are debating whether this means that large craters have faded away due to thermal evolution and relaxation of Callisto's icy crust, or whether the population of comets and asteroids that has impacted Callisto is lacking the larger projectiles that have struck the Earth and the other inner planets. *Courtesy NASA*

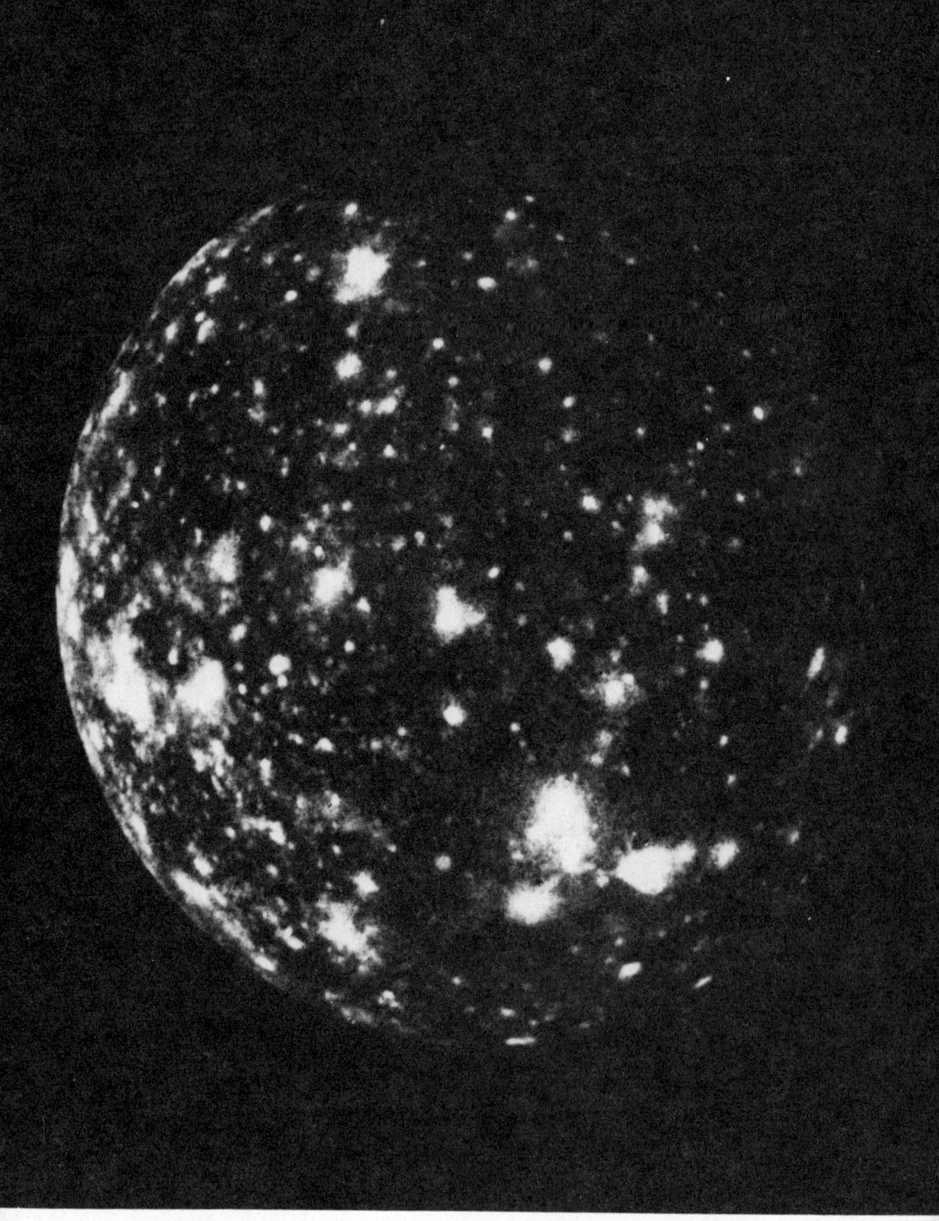

21. Bright, icy spots made by recent impacts are the only features visible on this global view of Jupiter's dark satellite, Callisto. *Courtesy NASA*

22. A gallery of small, irregular-shaped Saturnian moonlets. The separate pictures were all taken by Voyager 2 on the day of its closest approach to Saturn in August 1981. These icy objects range from ten to a couple of hundred kilometers across. Several of them help to shepherd Saturn's outer rings while others are so-called Trojan moons, sharing the orbits of two larger Saturnian satellites. *Courtesy NASA*

23. This Voyager 2 picture of Saturn's giant moon, Titan, shows a bland world enshrouded in atmospheric haze layers. Although geologists were disappointed that the hazes obscured Titan's surface, instruments aboard Voyager 1 were able to penetrate the atmosphere and measure conditions all the way to the surface of this weird world. *Courtesy NASA*

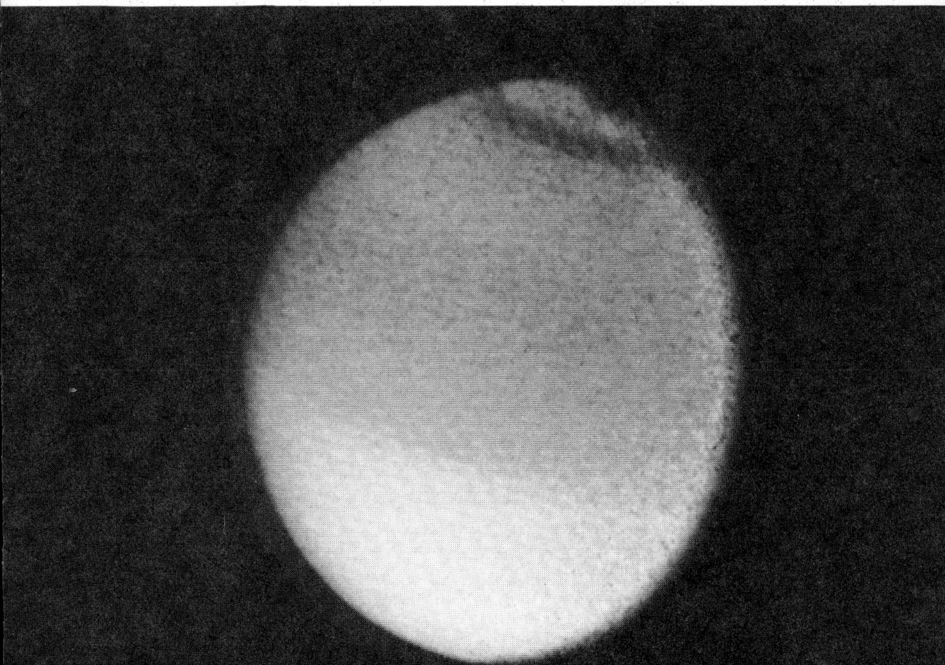

24, 25. Two different faces of Saturn's inner moon Mimas are revealed in these Voyager 1 portraits. One side is peppered with craters (below) and the other side less so (above). But the large crater with the prominent central mountain on the less-cratered side of Mimas is one of the largest known craters, in proportion to the size of the body on which it formed, in the solar system. *Courtesy NASA*

5

An Inner Planet Revealed

The opportunity to add a whole new planet to our base of knowledge about the terrestrial planets has been of extraordinary importance, the intellectual implications of which will continue to develop over succeeding years.
—Bruce C. Murray, 1975

It is often repeated that Nicolaus Copernicus never saw the planet Mercury. The tale is apocryphal, but of the five "wandering stars" known to the ancients the winged messenger Mercury is certainly the most elusive—a faint morning or evening star playing tag with the sun and never visible in a completely dark sky. Now geologists study Mercurian cliffs, ridges, and craters on aerial photographs transmitted from a spacecraft that itself played tag with Mercury. They are deciphering the history of this superficially moonlike world for which a decade ago we lacked even a crude map.

The spasmodic history of our learning about Mercury exemplifies the scientific method, its mistakes and its accomplishments, and especially the power of modern technology. Answers to some perplexing planetary riddles are being uncovered by diligent analysis of reconnaissance data beamed back to Earth by Mariner 10. Yet some of the most fundamental facts about the planet were uncovered years before any spacecraft went near it, by Earth-based radar techniques that literally reached out and touched Mercury.

As schoolchildren, we all learned a few salient facts about the smallest planet, the one nearest the sun. First, tidal forces pulled one face of Mercury always sunward, so that its day exactly equaled its year of 88 Earth days. Naturally, the sunlit side was very hot and the dark side very cold. We learned, in fact, that Mercury was simultaneously the hottest and the coldest place in the solar system—a planetary baked Alaska. We may have heard of a tenuous atmosphere on this inhospitable planet and seen maps showing man-in-the-moon-like dark patches on Mercury's sunward hemisphere. These were the facts, undisputed well into the 1960s. The French astronomer Audouin Dollfus went so far as to assert that Mercury's rotation period was known to be 87.969 days to a precision of 1 part in 10,000.

But scientific facts, like so many others, are evanescent. To be sure, Mercury is nearest the sun, but although it is tinier than some moons of the outer planets, the distant double planet Pluto is still smaller. And everything else we were taught about Mercury was sheer fiction. It is searingly hot at high noon on Mercury's equator, but not so hellish as practically everywhere on the surface of Venus. Nor does it ever get so cold on Mercury as in the outer solar system. The maps of Mercury's sunlit side were illusionary since there is no permanent sunlit side; Mercury does not keep the same face to the sun, but rather rotates once every 59 days or so. Nor is there even a tenuous atmosphere.

What went awry? It is a story of human fallibility, for Nature simply fooled us. It all started a century ago with an Italian astronomer, G. V. Schiaparelli, at the Royal Observatory in Milan. He is famous—or infamous—for "discovering" the nonexistent canals on Mars in 1877. From that developed the conception of the Red Planet as the harborer of life, indeed of civilization. Should Mars turn out to have always been devoid of intelligent beings, just as today it is surely devoid of encircling canals, the twentieth-century view of Mars inspired by Schiaparelli's modest observations will nevertheless endure in literature and history. But memory of his article "Sulla rotazione di Mercurio" ("On the rotation of Mercury"), published in 1890, will fade much sooner. And it is just as well, for if his observations of Mercury were conservative and his reasoning generally sound, they were also insufficiently perfect.

After tentatively observing the small rosy planet in 1881, Schiaparelli decided to make a regular study of it. He was the quintessential astronomer, peering at a twinkling planet through a long telescope and recording his observations in a diary, translated and paraphrased here: "I have observed Mercury in the telescope several hundred times and on more than 150 days it has been possible to see some spots on it, or at least something worth noticing. I have made about 150 drawings of it, which however are of uneven quality. But all more or less have contributed to the results of the present study."

Schiaparelli's logic in determining Mercury's rotation period from the positions of the spots had three elements which he conveniently numbered I, II, and III: "I. Observing Mercury on two consecutive days at the same hour . . . one sees the same spots . . . occupying approximately the same places on the apparent disk. . . . Of all the facts concerning the rotation of Mercury, this one is most manifest and the oldest known."

The German observer Johann Schroeter had made the same observation early in the nineteenth century. He had opted for the simple interpretation that Mercury rotates in 24 hours, just like the Earth. Schiaparelli pointed out two equally plausible alternatives. First, Mercury could make two or more complete rotations in 24 hours and present the same configuration of spots at precise 24-hour intervals. Or, it could rotate so slowly that the spots would not shift noticeably from day to day. Schiaparelli settled this question: "II. But observing the planet several times in the course of the same day at intervals of several hours, one still finds that its appearance is not changed. And the same is true when one repeats the observations on two consecutive days but at notably different hours, so that the interval is significantly greater or less than 24 hours. This fact is no less obvious than the preceding and is in open contradiction with the rotation of Schroeter. . . . Mercury rotates in neither a day nor a fraction of a day, but rather very slowly."

But with exactly what period? Schiaparelli's third and clinching observation was that even from year to year, the spots seemed to be reasonably fixed with respect to the boundary that divides Mercury's sunlit hemisphere from its night side. Only one conclusion seemed plausible; as Schiaparelli wrote, "The ensemble of these facts . . . shows that Mercury turns around the sun nearly in the same fashion as the moon around the Earth and Iapetus around Saturn, generally presenting to the sun (but with some oscillations) always the same hemisphere of its surface."

From the days of Kepler and Galileo, analogy has played a major role in planetary science. It was natural for Schroeter to have supposed Mercury's period to be 24 hours, like the Earth's, and just as natural for Schiaparelli to see a lunar analog once he had disproven Schroeter's result. Moreover, the reason for the moon's behavior had been proven mathematically just a decade earlier by Sir George Darwin, son of the evolutionist; the theory seemed applicable to Mercury as well.

Schiaparelli's penchant for thinking of Mercury in Earth's terms made him overlook hints in his own diary that otherwise might have seeded doubts. First were his oscillations in positions of Mercury's spots; these he managed to ascribe to a well-known lunar phenomenon that would be greatly magnified for Mercury: libration. Kepler had shown that a planet in an elongated orbit, such as Mercury's, moves around the sun at an uneven rate. Yet a planet's spin about its axis is uniform, so the rotation gets ahead of, or lags behind, the day-night boundary, depending upon whether the planet is moving slower or faster in its orbit. A more obvious hint was that Schiaparelli's spots were "sometimes more visible and sometimes less." On occasion spots disappeared altogether, only to reappear a week later. Schiaparelli had a ready Earth analogy for explaining these mutations: "It does not seem too rash to suppose that [these effects are due to] more or less opaque condensations produced in the atmosphere of Mercury that, from afar, appear analogous to the appearance that the Earth's atmosphere must present from a similar distance."

Later, during the 1920s, the great observer E. M. Antoniadi would likewise rely on veils and clouds to explain discrepancies between Mercury's appearance and his expectations based on the 88-day rotation period. Neither Schiaparelli nor Antoniadi considered that perhaps the spots, rather than being veiled, simply weren't there, having rotated around the planet. Still, the inconsistencies were infrequent and Schiaparelli can be forgiven for failing to attend to them. Usually Mercury does appear to behave just the way Schiaparelli reported—at least when astronomers are looking at it. And there's the rub! For Mercury cannot be kept under constant surveillance. It can be seen well only relatively briefly, during the six times a year it swings away from the sun. Three times it swings east of the sun and can be seen in our evening sky, and three times west into the predawn sky. From northern latitudes only two of these—an evening apparition in the spring and a morning one in autumn—provide superior views; at other times the image of Mercury is distorted by the turbulent, hazy air near the horizon. The interval between two such opportunities to observe Mercury's morning side (during evening apparitions) is just about 348 days, the same interval as between successive times when its evening side is favorably displayed.

Anyone who has seen wagon wheels appear to turn backward in a motion picture is aware of the illusions possible when a rotating object is viewed intermittently. This stroboscopic effect is as relevant to Mercury viewed every 348 days as to wheel spokes viewed each 1/24 second. The stroboscopic illusion works only if the rotating object has a peroid very nearly two, three, or some other integral number of times the viewing interval. A filmed stagecoach's wheels misbehave only when the coach moves just so fast; most often the wheel spokes are a blur. It just so happens that exactly six of Mercury's 58.65-day rotation periods take 352 days, just 4 days longer than the 348-day viewing interval. Since exactly four of the supposed 88-day periods also take 352 days, it is easy to imagine that Schiaparelli's observations would be equally consistent with the true period and the lunar analog period he adopted.

Schiaparelli also overlooked other periods that are consistent with his observation III. They are other whole-number fractions of the viewing interval (348/3 = 116; 348/5 = 70; 348/7 = 50; etc.). The 50-day period was a particularly serious omission, for with such a period Mercury would not only present the same face to Earth each favorable apparition but at every apparition, just as for the 88-day period Schiaparelli chose.

One further numerical coincidence helped fool Mercury observers. The 348-day interval between every third apparition is roughly equal to one Earth year. But it is not exactly 365 days, so the favorable spring apparitions occur about two weeks earlier each succeeding year; after 6 years or so they are no longer favorable spring apparitions but become mediocre winter apparitions, and eventually unfavorable autumn ones. Schiaparelli quit studying Mercury after 7 years, before the shifting apparitions brought new spots into view. And E. M. Antoniadi, who wrote the definitive book on Mercury, happened to observe only during the few years 1924 to 1929, when Mercury was, by chance, once again exhibiting the same faces during favorable apparitions that were studied by Schiaparelli; so there is little wonder that he confirmed Schiaparelli's result.

There is no reason at all for Mercury's true rotation period to be linked in any way to periods of the planet's visibility from Earth, so astronomers accepted the imperfect logic of Schiaparelli and Antoniadi. There was the lingering question of how this tiny, hot world could keep an atmosphere, required for the existence of reported veils, from evaporating into space. But Audouin Dollfus came to the rescue in the 1950s when he claimed—fallaciously, it turns out—that some properties of polarized light reflected from Mercury indeed proved the existence of a thin atmosphere.

We entered the Space Age secure in our portrait of the innermost planet. Complacency gave way to amorphous uneasiness in the early 1960s when improved radio-astronomical technology permitted detection of radio emissions from Mercury. A warm body radiates at all wavelengths, mostly in the "thermal infrared," but also at very long wavelengths called "radio." Mercury was a thin crescent and presented mainly its dark side to the radio telescope. But the surprisingly strong radiation revealed that Mercury's night side was warmer than the perpetually frigid temperatures expected. Mysteriously, the warmth from the sunlit side seemed to be leaking around, or through, the planet. Nobody considered that the supposed eternally dark side might have been basking in the broiling heat of the sun only a couple of months earlier.

The clincher came in 1965 from radar technology. A natural amphitheater in the Puerto Rican hills at Arecibo had been filled with a giant concave radar dish wider than three football fields strung end to end. With its immensely powerful transmitter, the dish focused a pure-note radio beam toward Mercury when the planet was overhead. Like a ripple in a pond, the pulse spread away from Earth, but traveling at the speed of light. A few minutes later, when the expanding shell-shaped beam swept past Mercury, the tiny portion of it that was intercepted by the planet was reflected back, just as a post sticking from a pond generates its own ripple when another ripple passes by. As the original pulse expanded infinitely toward the stars, Mercury's own weak echo sped backward through interplanetary space, again at the speed of light. Minutes later it flashed noiselessly past the Earth. But one radio receiver was pointed and tuned to receive the one-millionth of a trillionth part of Mercury's expanding echo-bubble that it intercepted—the same monstrous Arecibo dish that had emitted the original pulse a quarter hour earlier. The power of the Arecibo transmitter is exceeded only by the sensitivity of its receiver. So exact was the reception of the infinitesimally weak echo that the radar astronomers working in Puerto Rico didn't merely detect it but could analyze bits of the echo to determine Mercury's rotation rate.

How does one measure spin rate from an echo? Imagine you are facing a cliff a mile away and you clap your hands. About 10 seconds later, if you listen carefully, you will hear the echo. Clap several times again, once a second. The 10-second delayed echoes faithfully return, once a second. But imagine that, rather improbably, the cliff is rushing toward you as you clap once a second. The first echo is delayed nearly the full 10 seconds, but the cliff is much closer by the time your next clap reaches it, so the echo returns on the heels of the first one. And the third echo returns even more quickly. So the echo frequency is much greater than the one-a-second clap frequency. Without even seeing the onrushing cliff, a person might apprehend the imminent catastrophe threatened by the peculiarly speeded-up echoes. Just as the frequency of echoes increases as the mountain advances, so the frequency, or pitch, of the echo of a pure note reflected from the advancing cliff would be raised.

Now imagine that, instead of a cliff, you stand before a fixed, but rapidly rotating, globe. If the globe spins clockwise, then its right-hand portions are rushing toward you, and its left-hand side is receding. The parts rushing toward or away from you, as the globe spins, are somewhat farther away than the front side of the globe that is moving right to left. Suppose you could isolate just the later echoes from the most distant parts of this globe. Those coming from the right-hand side would be raised in frequency, or pitch, just as from the onrushing cliff or like the whistle of an approaching locomotive. But the simultaneous echoes from the left-hand (receding) side would be lowered in pitch. The faster the globe spins, the greater would be the difference in pitch of the echoes from the approaching and receding parts.

Similarly, the pure-note pulse from the Arecibo radar was slightly modified in pitch by Mercury's spin. The spread in pitch found in Mercury's echo was greater than expected from a planet spinning once every 88 days. Probably the spin was somewhere between 54 and 64 days. More precise radar measurements have since pinpointed the value at between 58.4 and 58.9 days. The announcement of the radar results caused consternation among optical astronomers, who only then reanalyzed the older drawings and found the occasional discrepancies with the 88-day period. Measurements of the drawings and some improved telescopic photographs refined the period to 58.65 days, almost exactly two-thirds of the 88-day orbital period. Finally in March 1974 the Mariner 10 spacecraft hurtled past Mercury in an orbit that brought it back again six months later. Measurements of close-up photographs confirmed that, within a hundredth of a day, the rotation period is exactly two-thirds the orbital period.

Not until the two-thirds period was found did astrophysicists reexamine Sir George Darwin's original theory to learn why Mercury fails to keep one face sunward. The answer lies in the same unequal orbital velocities by which Schiaparelli tried to explain away apparent oscillations of Mercury's spots. When Mercury is closest to the sun in its elongated orbit, it zips around so fast that even the 59-day spin lags behind. So when Mercury is being tugged most strongly by the sun, it actually does keep one side more or less facing sunward. Those periodic pulls on Mercury's slightly bulging figure overcome the weaker attractions when Mercury is farther from the sun and hold it locked into a precise two-thirds spin. How often it is that theory lags behind empirical data!

Before the first spacecraft exploration, radar astronomers had learned still more about Mercury. By timing the echo delays, they measured Mercury's diameter to nearly one part in a thousand. Radar also helped to improve our estimate of Mercury's mass by measuring Mercury's influence on the positions of other planets. Dividing mass by volume, the radar astronomers calculated that Mercury is made of material denser than the rocks from which our Earth is made. Presumably Mercury contains a great deal of iron, the only cosmically abundant material of high atomic weight. Mercury's high density had been suspected for decades from crude earlier data, but radar established it beyond doubt. Radar also provided an estimate of the roughness of Mercury's soil (rougher than Venus's) and demonstrated the lack of any Mercurian continental-scale highlands and lowlands, such as exist on Earth and Mars.

It is not often appreciated that ingenious improvements to ground-based optical-, radar-, and radiotelescopes have more than once enabled astronomers to do from the ground what experts had thought, a few years earlier, could be done only by the vastly more expensive examination of a planet by spacecraft. There has even been a tendency for the popular press, and uninformed spacecraft experimenters themselves, to credit the Space Program with discoveries made months or years earlier by ground-based astronomers. For instance, it is widely written that Mariner 4 first showed that Audouin Dollfus's estimate of the atmospheric pressure on Mars was too high by a factor of 10. Yet the year before Mariner 4 reached Mars, University of Arizona astronomers Gerard Kuiper and Toby Owen had lowered Dollfus's estimate by a factor of 5 and were close to deriving the currently accepted value. However, the dramatic feats of radar astronomy in the mid-1960s could not be overlooked. The unanticipated coup of reorienting our conception of Mercury gave the technique a tremendous boost in prestige. The attendant funding enabled the Arecibo radar to be improved to the degree that for several years now it has been mapping the surface of Venus beneath that planet's thick clouds in nearly as fine detail as Mariner 10's beautiful photographs of Mercurian landscapes.

On February 3, 1970, scientists gathered at the California Institute of Technology to review what was then known about Mercury and to plan the exploration strategy for the Mariner spacecraft that was to be launched in November 1973. The National Aeronautics and Space Administration had already selected teams of scientists to build the various instruments that would study Mercury's magnetic field, its charged-particle environment, and its infrared and ultraviolet radiation, as well as the cameras that would transmit back the close-up pictures of Mercury's surface geology.

Imaging-camera experiments always seem to be first in NASA's scheme of spacecraft priorities. Some people feel we learn more about a planet from pictures than by studying it in unfamiliar wavelengths or measuring the planet's interaction with its interplanetary space environment. But that is prejudice arising from our human biology—the superiority of our vision to our other senses. Instrument-design experts have overcome this bias, so in response to more worldly pressures they often vie with each other for specific spacecraft and mission designs that will benefit their own experiment. In the end the imaging experiment usually wins, however, for taxpayers like pictures and little understand or appreciate infrared radiation, electron spectra, magnetic-field intensities, or other technical measurements.

Yet on February 3, 1970, the imaging experiment was in trouble. Most scientists agreed that useful measurements could be obtained from virtually all other instruments only if Mariner flew past the dark side of Mercury. For instance, space physicists could hope to understand the interaction between Mercury's magnetic field and the wind of protons and electrons streaming away from the sun only by passing through Mercury's wake. Pictures could not be taken of the dark side, of course, and sunlit parts of the planet would be visible only from great distances long before or after encounter. From so far away, the available television camera, identical to that flown to Mars on Mariner 9, would yield much fuzzier Mercury pictures than Earth-based telescopic views of the moon. Imaging Team members used the Caltech gathering to rally support for a film (rather than television) camera system that would yield sharp pictures even from far away. (Some participants, noticing some film company representatives in attendance, thought the lobbying effort was actually the underlying purpose for the whole conference.)

Predictably, the choice of a camera system for the Mariner 10 mission had little to do with science and much to do with sociology. Geologists hoped to get general coverage of the sunlit half of Mercury, plus highly magnified views of at least the more interesting localities. But there were many other constraints. Only a finite number of pictures can be transmitted accurately back to Earth; some transmission capacity had to be reserved for other instruments on board. The film camera system advocated by the Imaging Science Team would require costs not anticipated in the mission budget, and it seemed risky to go with a system never tried before. Moreover, it was doubted that a film system would be as "photometrically accurate" as a television system. Scientists would often like to know not just that this place is darker than that one, but indeed that it is, say, just 17 percent darker; from such data they can measure how much the ground slopes or how big are the particles that compose the soils. Accurate calibration of photographs is difficult and television systems are ostensibly more accurate.

That was a sore point, however, for all previous television-equipped spacecraft had returned pretty pictures that nonetheless were photometrically worthless. The worst problem occurred in 1965, when Mariner 4 returned pictures of Mars that were horribly washed out and were made presentable only by extensive computer processing. Andrew Young, a brilliant, nonconformist expert on photometry, incurred the displeasure of some officials at the Jet Propulsion Laboratory by suggesting bluntly that Mariner 4's problem resulted from negligence. Stray light from Mars flooded the vidicon through a hole in the camera; the hole was a defect in the design, but nobody had bothered to check for such off-axis light leaks before installing the instrument in the spacecraft.

The arguments for the film system by Bruce Murray and the Imaging Team were impressive, yet there were counterarguments for frugality and conservatism. An eventual compromise enabled Mariner 10 to fly to Mercury equipped with the tested and true Mariner 9 television, but also with a larger telephoto lens that permitted sharp pictures to be taken from the great distances required by the dark-side trajectory. Additional technological advances further improved picture quality to nearly the predictions for the film system. Given that our previous best views of Mercury showed nothing more than Schiaparelli's vague, darkish spots, it was inevitable that any close-up imagery would have thrilled the photogeologists. So as encounter date approached, the imaging-system debate was forgotten. Even the 1970 Mercury conference itself faded from memory, and a subsequent conference held on the same Caltech campus in June 1975 was dubbed the First International Colloquium on Mercury.

By early afternoon on Friday, March 29, 1974, the level of excitement had reached a feverish pitch in the catacombs of the buildings on the Jet Propulsion Laboratory campus, nestled against the San Gabriel Mountains north of Los Angeles. Over 148 million kilometers away, the Mariner 10 spacecraft was winging past Mercury at more than 11 kilometers per second, having been propelled there from its fling past Venus only 7 weeks earlier. Its signals were being received by a specially upgraded tracking antenna in the Mojave Desert and relayed in turn to the Mission Control Room at JPL. Television monitors scattered around the JPL campus flashed pictures of a landscape never before seen. Later, and into the night, further data streamed back to Earth from a spacecraft that was already leaving its target far behind. Measurements made while Mariner was behind Mercury, and thus invisible from Earth, had been recorded on tape and were played back shortly before sunrise on Saturday morning.

Finally, the hard but rewarding work of serious analysis was to begin, analysis that would take months, perhaps years. After hiding in the sun's glare for an eternity, Mercury had at last been exposed to scrutiny.

6

The Inside View

One of the principal unsolved problems of geophysics is the nature of the source of the terrestrial magnetic field. Our knowledge of the properties of the interior of the earth . . . tell[s] us that the source is a dynamo but they do not tell us how [it] works nor what is the driving energy source. Mercury, Jupiter, and Saturn have planetary magnetic fields [almost] certainly due to internal dynamo processes. The contrast between the properties of these magnetic fields and the terrestrial field together with the contrast in their interior properties should provide some basic clues to the operation of the geodynamo.

—*C. T. Russell, R. C. Elphic, and J. A. Slavin, 1980*

As human beings living on the surface of a planet, we are naturally most concerned about that surface: the land, the waters that flow on the land, and the air that blows across it. The outer reaches of our atmosphere produce the shimmering light shows we call the aurorae, but otherwise seem of little relevance. Although we are amused by Jules Verne's fantastic voyage to the center of the Earth, we derive economic benefit from only a very narrow zone of our planet, near its surface.

As we comprehend the complexity of modern civilization, we become more aware of the interrelation of ourselves and our greater environment. Our ecological concerns are no passing fad, despite the efforts of some wrong-headed politicians. Instead, they represent the awakening of our society's deep-seated realization that we must understand our environment so that we can control our ever-widening effects on it and ensure our survival. We are learning that the ozone layer in the stratosphere, far above the altitudes at which most aircraft fly, is essential to our survival, and yet it is endangered by our use of aerosol sprays. The geological readjustments deep in the Earth threaten the survival of our largest cities. Ice ages may result from a crucial interplay between processes far removed from the Earth's surface; mighty volcanic eruptions, which originate deep in our planet as drifting crustal plates plow beneath each other, throw volcanic dust high in the stratosphere where it is trapped for months, affecting the balance of solar energy.

In this chapter I discuss one remote part of a planet—the deep interior. But, paradoxically, measurements in that other remote zone, far above the surface and atmosphere, tell us most about the interiors of some planets. The interior of our own planet not only constitutes the vast bulk of its mass, but its influence on us is surprisingly great, so we must understand it. And comparisons of Earth's interior with interiors of other planets help test our ideas about our own planet.

Planetary interiors may seem so inaccessible as to be quite undecipherable. But that is not the case. Were Earth a totally static body with its interior never manifesting itself on the outside, we would know nothing but its volume calculated from the measured exterior dimensions. But the mass of the Earth's interior pulls on us from a distance—a fact as fundamental to our existence as the solar energy that sustains us. The life-supporting waters and vapors are thereby held close to the surface on which we too were permanently confined until the Space Age dawned. From this same essential gravitational pull we can measure the mass of the Earth. Since we know its volume, we can calculate the Earth's density, which is a fundamental constraint on the chemical composition of the interior.

We learn about some of the minerals that compose the Earth's interior from direct examination. I have on my desk a heavy, granular, greenish rock made of olivine crystals, a material believed to be a prime constituent of the Earth's mantle. It is no coincidence that I found this rock while walking on the slopes of Mount Hualalai on the island of Hawaii, one of the most active areas of volcanism. The veins of olivine-rich rocks in which diamonds are mined were formed deep in the Earth and were brought to the surface in several places by our planet's active geological forces.

The earthquakes that threaten us provide another way of exploring the Earth's interior. While the mighty shocks cause damage only in limited areas, seismographs sense weakened shocks from all over the world. From the delays between quakes and their detection around the globe, seismologists calculate propagation velocities that constrain the densities of the layers of the Earth's interior. The forms of the seismograms further reveal whether the material traversed by the wave is solid or liquid. It is an unambiguous discovery of seismology that the outer layer of the Earth's core is liquid; indeed it must be molten nickel-iron with some admixture of silicon or sulfur.

Yet another manifestation of the Earth's interior is our magnetic field. It is produced by the convective boiling of the Earth's liquid metallic core, although the actual process is not yet well understood. For centuries mankind's exploration of our planet's surface was guided by compasses that are directed by these subterranean forces. The field is so strong that some rocks have retained magnetization from previous epochs. It is from studies of such rocks that the paths of continents and sea floors have been traced around the globe. The recognition that the continents move leads us in turn to search the depths of our planet for the origins of their motions.

Deciphering the internal properties of the Earth from these and other surface manifestations is a tricky business. What progress has been made owes much to laboratory measurements of the properties of rocky materials at the high pressures and temperatures that characterize the Earth's interior. Laboratory techniques fail to reach the central pressures of nearly 4 million atmospheres and temperatures of over 4,000 K (7,000° F), so theoretical physics and cosmochemistry must provide still further insights into the total picture.

Planets are like living creatures. They are born, full of life and activity. They mature, consume energy, and settle into established ways. Finally they run down, become dormant, and die. On a human time scale planetary lives are virtually eternal. We see only a snapshot of each planet and can only surmise its evolution. True, we can measure (barely) the several-centimeter-per-year drift of the continents. But most of the Earth's dynamic activity that we can actually measure, such as the rate at which the land is washed into the sea (estimated from the sediment load of rivers), is superficial and transient. Such surface processes would soon stop were it not for the much longer lived internal activity of our planet and of the sun. Yet the snapshot of our planet is filled with evolutionary clues, ranging from fossils to chemical isotope ratios. Moreover, well-understood laws of physics and chemistry enable us to predict generally the future evolution of the Earth.

What is fundamental to a planet's life history, as for the universe in general, is matter (of which it is composed) and energy (which keeps it moving, or "living"). At birth a planet has a certain amount of matter; maybe a lot, like the Earth, or less, like the moon. Planets also began with different temperatures; some originated near the newly formed sun and were hot, while others farther away were colder. (There were other even more important contributions to the early temperatures of the planets, which I have discussed before.)

Eventually the planets began to cool, to lose their heat. Heat can be moved in three ways. First, it can be directly radiated as light or infrared radiation. That is how one gets hot sitting before a fire, or sunburned by a star 93 million miles away. But just a sheet of cardboard shuts out light, so it is easy to believe that the Earth's interior hardly relies predominantly on radiation to move its heat out to the surface through thousands of kilometers of cold, opaque, solid rock. Radiation is of course an important way for surface heat to be transported out through the atmosphere and into space.

In opaque, solid materials, heat flows by conduction between touching objects. This is how one is warmed by touching a warmer object. Heat slowly flowing from a warmer place to a colder place is a major way that planets cool off.

If temperature differences are great and the material is fluid, convection occurs. This is the actual transport of a warm hunk of material to a colder region. When the sun warms air near the ground on a hot summer day but the air above is very cold, the warm air is buoyed up into a billowing thundercloud and replaced by cold downdrafts at the edges of the storm. Over eons of time even the solid parts of the Earth's interior are plastically deformable and are slowly stirred by convection. Warmer parcels of the interior rise toward the surface and transport heat faster than by conduction alone. Some geophysicists believe that continents are conveyed on top of such slowly churning convective cells within the mantle. Hot lavas pouring out onto the surface from deep inside the Earth are an extreme example of convection and graphic proof that the Earth is still "alive."

A planet's lifetime depends on its size. Just as a glacier lasts longer in the sun than does an ice cube, so a larger planet takes longer to cool than a smaller one. Hence the Earth should be alive long after the smaller asteroids and moon have cooled off, everything else being equal. But planets are wonderfully unequal. For instance, many moons of Jupiter and Saturn are made mostly of ice rather than rock. At temperatures at which rock is solid an icy body might be "molten" and fountaining tepid "lavas" we would call water.

Also, initial heat is not the only source of energy. For one thing, the sun shines and heats planetary surfaces. Jupiter's moon Io is heated feverishly by its tormented gravitational relationship with Jupiter and two neighboring moons. And the inner planets are heated chiefly by the decay of radioactive isotopes over their half-lives of several billion years. These isotopes were not equally distributed among the planets because of the variable composition of matter condensed from the cooling solar nebula. Moreover, they are not uniformly distributed within planets; because of their chemical affinities, they tend to float toward the surface of a rocky planet if it warms to the point at which iron begins to sink to the center.

Thus the fate of a planet depends not only on its size, composition, original temperature, and orbit, but also on the quantity and internal distribution of radioactives and on their half-lives. By using the best data on the thermal properties of rocks, geophysicists such as Sean Solomon of M.I.T. have been able to calculate how different parts of a planet become warmer or cooler. When it becomes warm enough for rocks to melt and iron to sink, Solomon moves all heat sources in his computer model into the upper regions of the planet, to simulate the migration of radioactives into the crust. He then uses thermal parameters for metallic iron in his mathematical representation of the core and those for rocks elsewhere. Gradually, as the computer churns away, the heat is calculated to migrate to the surface of the planet, where Solomon's computer code lets it radiate away. Finally, programmed by their half-lives, the radioactive heat sources waste away. The computer then predicts the inevitable fate of each planet's heat engine. Unless Solomon has overlooked something terribly important, the Earth is predestined to cool and die several billion years from now.

Internal heat in our planet has been essential for all that concerns us who live on the surface. The atmosphere and oceans have been outgassed from inside our planet by the heat, and heat-generated geological processes have shaped the Earth's crust. The internal death of the Earth need not doom life on the Earth's surface, however. After all, our dominant energy is from the sun, which will still be shining, perhaps more brightly than ever. But the ecological effects of the Earth's internal death will be profound and are not easily predicted. Volcanism will stop and the continents will cease to drift. Mountains will no longer form and if our atmosphere were unchanged—a very dubious proposition—the rains would wash the land into the sea. The eventual fate of Earth depends on the finite supply of fuel for the thermonuclear reactions within the sun. The sun will evolve and grow larger, then ultimately will die itself. If it does not engulf and destroy the Earth in the process, our planet will be left as a cold cinder drifting in the blackness of space.

For our knowledge of the Earth's interior we rely on indirect evidence, much of which has been accessible because of the extreme geological activity of our planet. The churning motions constantly signal our seismic listening posts, toss interior rocks onto the surface, and generate a strong magnetic field. How much more difficult it must seem to learn about the interiors of less active planets, or of the moon, whose ancient cratered surface testifies to the relative tranquillity of its interior. It is as though an amateur spy in a cheap motel were trying to learn about the occupants of neighboring rooms. From one room come shrieks of laughter, loud music, and the sounds of smashing glass. Perhaps a fist crashes through the thin dividing walls. Our supersleuth would have no difficulty guessing that a drunken party was in progress and, by listening to the raucous noise, might even learn the names and personalities of those attending. But from the other room there is no noise at all. Is the room occupied by a sleeping man? Or a woman? Or a dead body? Or is it a broom closet and not a bedroom at all?

The problem of learning about other planetary interiors is exacerbated by the brief and quite superficial nature of our exploratory ventures into space. The seismic stations on the moon, with which we listened to the occasional, weak moonquakes, were expensive and pitifully few. Not even a mechanical lander has been on Mercury. The two Soviet Venus explorers that took pictures from remote locations on that planet's forbidding surface in autumn 1975 survived the heat for only an hour. Thus we must study the relatively passive interiors of many terrestrial planets from afar, without benefit of touch.

Still, the advances in comprehension of the Earth, primarily in the 25 years since the International Geophysical "Year" (July 1957–December 1958), have given geophysicists sufficient background to infer the nature of other planetary interiors given only data sensed remotely and returned from spacecraft flying past.

Consider again Mariner 10's reconnaissance of the planet Mercury in 1974. Newspapers printed photos of Mercury that reminded everyone, astrogeologists included, of the pockmarked surface of the moon. I have discussed in earlier chapters how these monotonous crater fields hold important implications for the rest of the solar system. Yet the most significant results from Mariner 10 are not primarily from the surface pictures but concern the interior of Mercury as it has been deduced from a synthesis of several Mariner 10 experiments, theoretical calculations, and the few facts that were known before.

Even the most passive planetary interior reveals itself by its gravitational effect on the motions of neighboring planets and asteroids. So it was long suspected, and more recently known for sure, that Mercury is quite dense. Almost certainly it is composed of a large amount of iron, nearly two-thirds by mass. This contrasts with about 35 percent iron for the Earth. The ordinary chondrite types of stony meteorites, which are thought to be unmodified condensates from the primordial solar nebula, have only 20 to 25 percent iron.

A fundamental question is whether the iron is uniformly distributed throughout Mercury as it is in chondritic meteorites or is concentrated in a central metallic core as in the Earth. If the iron were uniformly distributed, then Mercury could be an unmodified, ancient aggregate of condensates that were simply more iron-rich than in the portion of the solar system where the chondrites formed. Even with an iron core, Mercury could be unmodified if it accreted so fast that it grew layer upon layer as successive minerals condensed from the more slowly cooling solar nebula, beginning with iron. According to the prevailing models of planetary formation, however, Mercury should have formed from homogeneous material, similar to (though more iron-rich than) the ordinary chondrite meteorites. If Mercury has a core, then the planet must have heated and partially melted, permitting the heavy iron to sink to the center of the planet. Such core formation, believed to have occurred on Earth early in its life, would virtually have turned the planet inside out and had a catastrophic effect on its surface.

Mariner 10 revealed evidence for the probable existence of a core in Mercury, but it also raised further questions. The most convincing evidence came, ironically, from instruments that measured the most superficial aspects of Mercury: the nature of interplanetary space near the planet. The three teams of scientists running the plasma science experiment, the magnetometer, and the charged-particle telescope were all surprised at the implications of their data: Mercury has a magnetic field! While it has less than 1 percent the strength of the Earth's magnetic field, it is still substantial, contrasting markedly with the virtual absence of fields around Mars and Venus.

The discovery was unexpected, for while a molten core is thought to be a prerequisite for a planetary magnetic field, rapid rotation was thought to be required as well. Mercury's languishing 59 day spin hardly seemed to qualify, and Mariner researchers had expected to study Mercury's passive interaction with the interplanetary fields and associated charged solar particles streaming past the planet. Such passive interactions, carefully analyzed for the moon by a series of lunar satellites and later for Venus by the Pioneer orbiter, are sometimes complex and superficially seem to mimic an active, intrinsic magnetic field. Thus, the experimenters were at first uncertain of their discovery. After all, there is an awfully large volume of space around Mercury and on March 29, 1974, Mariner 10 made only a single traverse of it.

But by a fortunate circumstance it was possible for Mariner 10 to pass through Mercury's field again a year later. Although not recognized when the Mariner Venus–Mercury mission was first planned, the spacecraft's path was a peculiar one that brought it back near Mercury every 6 months. During the second encounter, slight course adjustments were made so that Mariner would pass high over sunlit parts of the planet, outside the magnetosphere, to get wide photographic coverage of the sunlit south polar regions that were poorly observed on the first pass. But there was a third encounter, in March 1975, which was once again aimed close to the night side of the planet, though closer to the pole than the first pass. Spacecraft instruments don't function forever; the third Mercury encounter proved the last for Mariner 10. Yet the third encounter provided a second cut through Mercury's magnetosphere that was sufficient to confirm the expectations from the first pass that Mercury indeed has an intrinsic magnetic field.

An intrinsic magnetic field isn't necessarily generated by a currently molten core, however. After all, a bar magnet doesn't have a molten center, but nobody has deduced how Mercury's presumably iron-depleted rocks might have been sufficiently magnetized to produce the observed field. One current hypothesis is that, despite Mercury's slow rotation, its field is being generated in the same rather mysterious way that the Earth's field is. That implies not only that Mercury has an electrically conducting iron core, but also that the core is at least partially molten.

We could also infer the presence of a core from examining Mercury's surface composition. This might seem paradoxical at first, but if the iron required by Mercury's density were absent from the surface, then it must have sunk at least partway toward the core. In Chapter 4 I described how Earth-based astronomers measure the composition of surface rocks on asteroids and other distant bodies. For Mercury, however, the techniques fail to provide a clear clue to the percentage of iron present.

An indirect clue about surface composition comes from interpretation of the television pictures of Mercury's widespread, so-called intercrater plains. In many respects they resemble the moon's lightly cratered volcanic maria, so Mariner 10 geologists think they too were once composed of molten rock. And iron will surely sink beneath the areas where molten rock originates. Expanses of volcanic plains certainly would suggest that Mercury segregated into a metallic core and rocky mantle. But this nifty argument for core formation, which is the current consensus, is not totally convincing. It bears an uncanny resemblance to arguments proposed many years ago that some bright highland plains on the moon were volcanic. When Apollo 16 astronauts brought back rocks from such a plain, they turned out to be impact-smashed crustal rocks from—or at least produced by material ejected from—the giant multiring impact basins on the moon. They were not volcanic at all, to the embarrassment of U.S. Geological Survey geologists who had confidently mapped the bright volcanic plains. Mercury's intercrater plains are more widespread than the moon's, while potential source basins are only one-third as common. In fact, even if basins were numerous, Mercury's higher gravity would keep basin material from being hurled so far as on the moon. Still, there are no obvious volcanoes or other proof that Mercury's plains are volcanic, so they may say nothing about a core within that planet.

Another trait of Mercury's topography provides more secure

evidence for a molten stage and core formation on the planet. One feature on Mercury is virtually absent from the moon: the so-called lobate scarps, or cliffs, which are scattered all over Mercury and range from tens to hundreds of kilometers in length. Several are shown in Illustration 4. They have the special shapes and forms of particular scarps on the Earth that form when immense compressional forces thrust one block of land on top of another. In fact, Mercury's crust resembles rather closely the wrinkled skin of a piece of fruit that has dried out and shrunk inside.

What could have caused Mercury's interior to shrink, leaving a crumpled surface? The only plausible answer is that the interior was once warmer and contracted upon cooling. Indeed, Sean Solomon has calculated that the cooling of normal rocky materials in a Mercurian mantle alone would account for the shrinking indicated by the size and number of lobate scarps on Mercury. Since most large craters have been wrinkled and yet a few craters formed on top of some wrinkles, the contraction of the mantle must have begun near the end of the period of heavy cratering, presumably a long time ago. Solomon has further shown that if Mercury's core had also cooled and solidified, much greater contraction should have resulted, crumpling Mercury's surface much more than is observed. He concludes that Mercury has yet to contract and that much of its core must still be molten, as is suggested also, of course, by the presence of the magnetic field.

The logic seems obvious. Yet if it is correct, there must be something dreadfully wrong with our understanding of the chemical composition of planets, or of the thermal evolution of planets, or with the cratering chronology for Mercury. Solomon has used his computer programs to trace the evolution of Mercury's temperature, using current hypotheses for the chemical composition of Mercury and sources for the heat that has melted it. The surprising answer is that, in the absence of some additional source of heat early in Mercury's history, Mercury's expected complement of radioactive uranium, thorium, and potassium should not have heated the planet sufficiently for the iron to sink inward for 1½ billion years. That is long past the popularly hypothesized cratering cataclysm. Yet once the core formed, it must have cooled rapidly, according to Solomon's computer runs, and should have solidified long before now. So the straightforward model yields a core that forms too late to explain the wrinkled craters, yet solidifies too early to explain the magnetic field and lack of still more wrinkles.

I have mentioned in other chapters the independent evidence afforded by meteorites and the asteroids for an early source of heat for those bodies besides long-lived radioactive decay. Such early heating of Mercury would permit its core to have been formed earlier. Thus the association of wrinkles and craters may provide a further clue that early heating was a widespread occurrence in the solar system and not a peculiarity of the evolution of just one, or a few, planets.

More difficult to explain is the apparent failure of Mercury's core to solidify. There must be a large continuing heat source in Mercury's interior. Possibly the repetitive stretching and compression of the planet by the ever-changing solar tides raised in it as it swings close to the sun and away again contribute some heat. Or perhaps Mercury has more radioactive isotopes decaying within it than prevailing cosmochemical models predict. Maybe Mercury's crustal rocks are more iron-rich than we expect, sufficient to retain remanent magnetization from an ancient core dynamo now dead. Maybe we even misunderstand the overall time scales for Mercury's development as inferred from craters and their relationships to the great wrinkled cliffs.

The study of Mercury's interior exemplifies the impetus given to planetary science by new observations. Theoreticians who worry about how magnetic fields are produced in planets have had to rethink models that depended upon a rapid planetary spin. They have been forced to ask afresh why slowly spinning bodies like Venus and the moon lack such fields. Chemists are spurred on by the challenging possibility that their models for Mercury's composition may be more imperfect than they thought. And so on. Scientists who had never considered Mercury at all may now find their esoteric laboratory or theoretical work to be the key to rendering compatible the seemingly conflicting constraints.

Even our perception of the Earth's interior and its magnetic field may require revision. We cannot foresee how important, or unimportant, these changes may be for our day-to-day lives. One example illustrates the potential importance for our past. Rocks that record the strength and direction of the Earth's magnetic field show that it changes direction every quarter of a million years or so. The north magnetic pole is in the Antarctic, and then it flips back. It is not known how long it takes the field to flip, only that it is much shorter than the duration between flips. These traumatic events in our planet's magnetic life could be very long compared with a human lifetime, and should the magnetic field somehow turn off during flips, there could be dire consequences. After all, our magnetosphere controls the influx of charged particles from interplanetary and intergalactic space. There could be modifications of our lower atmosphere and dangers for living things if this protective regime were disrupted. Indeed, there are hints in the fossil record that some extinctions of certain species may have coincided with magnetic-field reversals. On the other hand, a thorough understanding of these processes, achieved through comparisons with the magnetic fields of other planets, may demonstrate that the flips are quite harmless.

7

The Vapors of Venus and Other Gassy Envelopes

The water gleamed, the sky burned with gold, but all was rich and dim, and his eyes fed upon it undazzled and unaching. The very names of green and gold, which he used perforce in describing the scene, are too harsh for the tenderness, the muted iridescence, of that warm, maternal, delicately gorgeous world.

—*C. S. Lewis,* Perelandra

Breathe deeply. Think of the fluid envelopes that surround the Earth—the air and the seas. Their miraculous properties shape the variegated environments that have provided the ingredients for the origin and sustenance of life. Without the atmosphere and the hydrosphere the Earth's surface would be very simple, despite all the subterranean churnings. Cratering impacts and occasional quakes would shake the ground, as on the moon, and only gravity would pull molten or fragmented rock downhill. The minerals would be few and life processes impossible.

But with gaseous and liquid coatings the Earth's surface is transformed into a veritable chemistry lab of cycling reactants and end products. The fluidity permits rapid migrations; sediments and raindrops sink, bubbles and light gases rise, animals move about. Only partially transparent, the air and water absorb some of the sun's energy, transporting it far and wide in currents and storms. Powerful sparks flash through the sky, relieving charge imbalances. Rains flush the ground, exposing fresh rock to chemical attack. Massive deposits of sediments accumulate on the sea bottoms, eventually to be cycled through the Earth's high-pressure oven below and spewed forth again from volcanoes.

This rich, wet, and gaseous chemical feast made the origin of life feasible, indeed probable, on our planet. If the processes that created our gassy and liquid envelopes were unique to Earth, other rocky worlds would be monotonously moonlike. The possibilities for life on them would be remote. Yet visual astronomers of the nineteenth century never doubted the Earth-like traits of our neighbors in space. They had good evidence for vapors on the other worlds. They couldn't appreciate, from watching dust clouds drift across Mars's surface or scrutinizing the cloudy veil of Venus, how terribly un-Earth-like those extraterrestrial atmospheres were.

Astronomers have long known that Venus has an atmosphere. That was not an inference just from the brilliant and nearly featureless expanse of creamy white that always seems to shroud the planet. When Venus passes nearly between Earth and the sun—closer to us than any other planet comes—its thin crescent extends into a ring of light, entirely surrounding the planet. There can be no doubt: we are witnessing both dawn and twilight on our sister planet as sunlight is refracted through a substantial atmosphere. Beneath the shroud Venus has hidden her secrets from us and remained more mysterious than planets ten times more remote. Through brute force, however, modern technology has finally enabled us to penetrate the veil.

To understand Venus's atmosphere is to understand Venus and much about Earth as well. Venus is nearly the same size and density as Earth and is not that much closer to the sun. We say the barometric pressure is high at 30.5 and the temperature hot at 98°F (37°C or 310 K), while on our twin planet normal conditions are a pressure of 3,000 and a temperature of 875°F (740 K). Venus's clouds of sulfuric acid blanket a choking ocean of carbon dioxide, a hundred times as dense as our own atmosphere. At night the ground glows dully from the extreme heat; the days are gloomy beneath perpetual clouds.

Imaginative fiction writers and scientists alike found Venus's hidden surface fertile ground for speculation. Astrophysicist Fred Hoyle imagined oceans of oil on Venus that would have satiated our energy needs. C. S. Lewis portrayed a watery world. The first hint about conditions on Venus came in the mid-1950s when radio astronomers detected a strong emission from Venus. It took a while to prove the emission was due to a hot surface, rather than to something else. Subsequently, ground-based radar echoes from Venus showed that its day is very long, 243 Earth days, and that it spins from east to west, rather than in the opposite direction, like Earth and most other planets. Radar probing determined the diameter of the solid part of the planet and, in combination with other data, showed that the cloud tops are about 60 kilometers above the surface. As the power and sensitivity of radar observatories have improved, we have gotten increasingly detailed maps of canyons and other topography far beneath Venus's clouds. Earth-based optical astronomers contributed to our understanding of the structure and composition of Venus's uppermost atmospheric layers.

Meanwhile, numerous spacecraft have reconnoitered Venus. While American Mariners were content to study it from afar, the Soviet Union was dropping instruments into its atmosphere. The early Soviet experiments frequently quit operating prematurely, providing dramatic evidence of the unexpectedly hostile environment near the surface. In 1974 excellent pictures of Venus's cloud patterns were taken by Mariner 10 on its way to Mercury. In October 1975 the Russians achieved a major first: they radioed back pictures from the surface of Venus, showing plains strewn with oddly shaped rocks. Although they succumbed to the heat after an hour, Veneras 9 and 10 first clocked winds of 2 to 8 miles per hour in the soupy air. Above the cloud deck the American Mariners found that Venus lacks a magnetic field; it interacts passively with the solar wind, just like a comet.

In December 1978 NASA put Pioneer Venus into orbit about our sister planet and dropped five probes into its atmosphere.

During the same month the Soviet Union sent Venera probes 11 and 12 to Venus. At last we have enough data in hand so that chemists, physicists, and mathematicians can employ their brain power (computerized and natural) to understand how and why the Venusian atmosphere behaves as it does. Some intelligent speculations are even emerging about why the volatile envelopes of the twin planets, Venus and Earth, have evolved so differently.

Why is Venus so hot? It is not simply because it is near the sun. Mercury is closer, yet cooler. Actually Venus should be comfortably cool. Even a small telescope reveals the brilliance of Venus's clouds; so much sunlight is reflected that little must be absorbed. Just as a house with a white roof and walls is much cooler on a sunny day than a black-shingled house, so Venus might be expected to be cool beneath its clouds.

The origin of the planet's heat has beeen controversial for two decades, the clouds themselves figuring prominently in the debate. Unless Venus were nearly molten due to heat generated within the planet—and that seems most unlikely—the clouds must behave differently from white paint. According to Carl Sagan, they must behave like the walls of a greenhouse. To visualize the greenhouse effect, imagine a greenhouse made of glass. Glass is transparent to light, so the sun shines through and heats up the ground and plants inside. They in turn radiate the heat back. But whereas the sun's 6,000 K surface radiates primarily visible light (it is no accident that our eyes are tuned to the sun's wavelengths), the reradiation by the room-temperature plants occurs at much longer, infrared wavelengths. Glass is opaque to infrared radiation, absorbs it, and reradiates it back to the plants. Since a little heat does leak out, because glass is not perfectly opaque to thermal radiation and also conducts heat, the temperature in the greenhouse eventually stops rising and stabilizes at a warm temperature.

It was recently pointed out to embarrassed meteorologists, who have debated the relevance to Venus of their greenhouse calculations, that this effect may not even be very important for greenhouses. Outside ground warmed by the sun heats adjacent air, which then floats upward to where the barometric pressure is less.

The air parcel expands, cools and settles into equilibrium. Meanwhile, at the ground the warmed air is replaced by cooler parcels from above. This process of stirring or convection, mentioned earlier in the book, warms upper regions and keeps the air near the ground from getting too hot. Air on the Earth begins to convect whenever the temperature begins to drop with altitude more quickly than about 6½°C per kilometer. So except in an inversion, when the upper air is relatively warm, convection maintains the 6½°C-per-kilometer temperature profile, which is why mountaintops are cool. The reason it is warmer inside than outside a greenhouse is mainly that the roof keeps the warmed-up inside air from floating away by convection. The greenhouse effect, although a perfectly valid physical principle, contributes only some of the additional heat inside a greenhouse.

There is no lid on Venus and its dense carbon dioxide is free to convect. The temperature profile of Venus's atmosphere ranges from 740 K at the ground to 250 K at the cloud tops. That is just the 8°C per kilometer expected for a convecting atmosphere of nearly pure CO_2. Can the combination of 100 Earth atmospheres of CO_2 plus the clouds plus whatever else is there serve as the hypothetical greenhouse glass? Is the Venusian atmosphere sufficiently transparent in the visible spectrum and opaque in the infrared to account for the furnacelike heat?

The small amount of carbon dioxide in our own air absorbs some radiation, making it difficult for astronomers to measure stars at some infrared wavelengths. The vast amount of CO_2 on Venus absorbs still more, but holes remain in CO_2's infrared spectrum allowing some radiation to leak out, keeping the temperatures from rising. Another absorber is necessary to maintain the heat. In the 1960s Carl Sagan advocated water in the form of vapor and ice clouds as the missing absorber. Although Earth-based spectra revealed little measurable water above Venus's clouds, Sagan calculated that a layer of water-ice clouds, with a humid atmosphere beneath, would yield sufficient infrared absorption to make the Venus greenhouse work.

Beyond being the preeminent popularizer of astronomy on television shows, the loquacious Sagan has been one of the most dynamic and influential planetary scientists since the early 1960s.

A man of vivid imagination, he keeps alive a wide variety of conceptions of planetary environments. By suggesting often outlandish alternatives and challenging traditionalists to disprove them, he inspired doubts about many accepted theories. Sagan's role has been essential for a healthy science because a bandwagon effect frequently leads to premature consensus among scientists before equally plausible alternatives have even been conceived, let alone rationally rejected. Carl Sagan regards his advocacy of unusual hypotheses as an intellectual sport, designed to keep his colleagues on their toes, and he claims to be dispassionate about the actual truth of his models.

But Sagan did not always take a detached view of the Venusian clouds, which were a major part of his serious research since his days as a graduate student. Long after most of his colleagues agreed that his once-accepted water-ice model for the clouds of Venus was incompatible with radio and polarimetric data, Sagan—with his agogic accent—continued to press his argument. Sagan watchers were forced to conclude that he actually believed in water-ice clouds (as most scientists believe in their own theories). Some even wondered if he also believed in his whimsical gas-bag creatures on Jupiter or polar bears on Mars! But no plausible alternative to water-ice clouds was proposed.

Then, in March 1973, at a meeting of the Division for Planetary Sciences of the American Astronomical Society, all the confusing data about Venus's cloudy veil began to fall into place. In a remarkable set of talks in a crowded auditorium in Tucson's Executive Inn, three scientists reported independent research that suggested the clouds were made of droplets of sulfuric acid. The first to think of sulfuric acid seems to have been Godfrey Sill, who works at the Lunar and Planetary Laboratory in Tucson. His arguments for sulfuric acid were bolstered by those of James Pollack, of NASA's Ames Research Center near Palo Alto, and of Andrew Young. Carl Sagan immediately agreed. But what about his earlier evidence for water? The Venusian clouds, although highly concentrated (at least 75 to 90 percent acid), are after all a *water* solution of sulfuric acid. The vapor pressure of water above such an acid cloud must be very low, which is why such tiny amounts are measured spectroscopically from Earth.

Now the search continued for water vapor *below* the clouds, since even carbon dioxide and sulfuric acid hazes combined cannot fully simulate the greenhouse glass.

A brief boost to Sagan's earlier view came when the Russians reported their early Veneras had measured quite a lot of water. Michael Belton, an astronomer at Kitt Peak National Observatory, gave some fellow astronomers a chemistry lesson at a 1974 Venus meeting in New York to help them evaluate the Venera result. He explained that on the Venera probe, the Venusian air was let into a double chamber separated by a pressure-sensitive membrane. One chamber contained a pellet of potassium hydroxide, which absorbed carbon dioxide. Afterwards the pressure difference between the chambers indicated how much carbon dioxide was in the Venusian air. Belton reminded his audience that Venera sampled air from within the sulfuric acid cloud, which was not known to exist when the Russians designed their spacecraft. Reliving his high school chemistry lab experiences, Belton produced a beaker of sulfuric acid. With a mischievous grin he remarked, "I don't think this has ever been done in the annals of planetary astronomy. Let's see what happens when we put potassium hydroxide in the sulfuric acid." With some understatement, the published conference proceedings record the result as, "Fizzle. Fume. Fizzle." So much for the Venera analyses of the composition of Venus's air!

In 1978 the Russians used a more sophisticated approach on their Veneras 11 and 12 and measured small but significant amounts of water vapor in the lower Venusian atmosphere. The concurrent American Pioneer measured still more water, which might even be enough for an isolated water cloud to form, but the result is considered less reliable. Both missions found that sulfur dioxide, which also can help block the escape of solar heat, is even more abundant than water vapor. Does sunlight really reach the surface of Venus to keep it hot? The Russians had been sufficiently dubious on this point that they had equipped the earlier Venera landers with floodlights to illuminate the scene for the cameras. They proved unnecessary. Pioneer Venus later determined that about 2½ percent of the sunlight filters down to the planet's surface.

Even after the invasion of Venus by spacecraft in December 1978, the best evidence about the composition of its clouds remains the Earth-based data. While no spacecraft instrument was designed to measure cloud chemistry directly, analyses are consistent with the presence of tiny droplets of concentrated sulfuric acid, probably contaminated slightly by other chemicals. The droplets pervade three stacked cloud decks, the lowest one based 47 kilometers above the surface. The inlet of one Pioneer instrument, designed to measure gas composition, unexpectedly became clogged as the probe parachuted through the 50-kilometer altitude. As it gradually cleared again, gas measurements indicated that the inlet was plugged by an actual evaporating droplet of sulfuric acid. (As we will see later, that one small cloud droplet proved to hold an important key for unlocking the secrets of Venus's past.)

Calculations of infrared blockage by all constituents of Venus's atmosphere now show that carbon dioxide is responsible for the lion's share of the greenhouse effect. But water vapor comes in second, partially vindicating Carl Sagan. Some blocking is also due to the sulfuric acid clouds and sulfur dioxide gas. It appears that the long-standing debates have been resolved at last. But if we can't even be sure that the greenhouse effect operates in greenhouses, perhaps we shouldn't be too dogmatic about Venus.

As Pioneer Venus answered some questions about our neighboring planet's cloud deck, it raised others. Lightning apparently was detected, which specialists always thought required large droplets or ice pellets to fall past each other in order to separate atmospheric charges. Yet strangely, the cloud particles on Venus come in three distinct sizes, *all* tiny. The largest particles are elongated crystals, which may be nitrosylsulfuric acid ($NOHSO_4$), according to Godfrey Sill. Such solid bits of acid may be produced by chemical reactions in the atmosphere of Venus that mimic the so-called Lead Chamber process used in the last century to manufacture sulfuric acid and fertilizers. When the process was poorly managed and inadequate water was injected into the chambers, chamber-crystals of $NOHSO_4$ were formed; Sill thinks this happens in the desiccated Venusian clouds. Even when monitored carefully, the process yielded sulfuric acid contaminated with dissolved $NOHSO_4$; such acid droplets are the same color as Venus. The middle-sized Venusian cloud particles, a couple thousandths of a millimeter in diameter, apparently are such contaminated sulfuric acid droplets. The crystal-and-droplet clouds are imbedded in a haze of still tinier particles, which may be the relatively nonvolatile anhydride residues that Godfrey Sill thinks would remain after evaporation of the settling larger cloud particles. The haze may serve as the cores, or condensation nuclei, about which the larger droplets and crystals form. It remains to be seen whether the historical manufacturing process will provide us with still more insight; for example, about why there may be lightning on Venus.

While Godfrey Sill has been learning about Venusian clouds by reading old handbooks on industrial chemistry, Venus research has had its own practical spin-offs for us on Earth. Sulfuric acid, after all, contributes to the eye-stinging quality of smog in cities where high-sulfur fuels are burned, and there is more than a casual analogy between the Venus clouds and smog. Before sulfuric acid was identified on Venus, several chlorinated chemicals were proposed as alternatives to Carl Sagan's controversial water clouds. A number of planetary physicists made theoretical studies of the behavior of hypothetical chlorine compounds in the upper atmosphere of Venus. Shortly after their work became irrelevant for Venus, their results were applied to a more urgent problem when it was realized that chlorinated aerosol-spray propellants rise into the Earth's upper atmosphere and consume the protective ozone layer.

Now we think we know what the Venusian atmosphere is made of and why it is hot. But what is the weather like on Venus? Are there winds and storms that make Venus's atmosphere a participant in the activity and evolution of the planet? On Earth we depend on the static properties of our air—the oxygen and the temperate climate. But equally essential for us are rains, alternating with sunny days, and the winds that blow, scattering seeds and dispersing smog. We must also understand such damaging weather phenomena as hurricanes, hailstorms, and droughts.

It may seem presumptuous to analyze the weather on Venus

when our own is so poorly understood. Yet it is for very practical reasons that we must study Venus. The failures of the weather bureau to warn us of last week's blizzard or of the thunderstorm that washed out the company picnic result from the lack of theoretical understanding of instabilities in fluid flows. These problems have been attacked by ingenious mathematicians faced with extremely complex equations and too few data from the only laboratory at hand, the Earth's atmosphere. Fast and powerful computers can precisely model the evolution of storms, but these numerical storms develop more slowly than the real storms outside the window. Another terrible complication in weather prediction is that our atmosphere is so unstable that the tiniest swirl of leaves can grow into a mighty storm in just a couple of weeks. It would be impossible to have a network of weather stations that could keep track of every such whirlwind.

In order to simplify the equations and make shortcuts for the computers, it is necessary for meteorologists to learn better how and why the atmosphere churns about as it does. They know that density is important, as are the vertical temperature profile and gravity. The heating is fundamental since it drives the atmospheric "engine." Also crucial is the rate at which a planet spins. Fluid dynamicists think they understand how some of these factors are related to the large-scale circulation. They have tried to test their ideas by dropping dyes into rotating dishpans of water and tracing the motions of the dyes. But the Earth is vastly larger, and there are many characteristics of its oceans and atmospheres that can hardly be modeled in a laboratory. Just imagine how you would make and hold together a free spherical shell of fluid in a lab!

So Venus, Mars, Jupiter, and other enveloped planets are natural laboratories for testing theories of the general circulation of our own atmosphere. The atmosphere of Venus is much thicker than ours, that of Mars much thinner. The air on Mars is warmed directly by the sunlit ground (except during planetwide dust storms), while Jupiter's clouds are warmed both from the outside by the sun and from the inside by the giant planet's own internal heat. Venus spins much more slowly than the Earth, Jupiter much faster. Mars has less gravity than the Earth, Venus the same, and Jupiter much more. A whole generation of theoretical meteorologists are testing their models for terrestrial meteorology by applying them to the vast quantity of spacecraft data from these other worlds.

Nearly two and a half centuries ago English meteorologist George Hadley presented a paper before the Royal Society of London in which he proposed an explanation for the trade winds. The sun heats the tropics much more than the poles, he noted, yet the poles are not all that cold. Hadley suggested that tropical air rises and flows toward the poles, forming a circulating cell with an undertow of cool polar air returning to the tropics at low levels. This "Hadley cell" formed a major element in models of the Earth's circulation long after it was realized that twisting of flows by the fast rotation of the Earth disrupted single-celled transport of heat between the tropics and the poles.

In the mid-1970s it became clear that Hadley's picture of the Earth's circulation might be more applicable to slowly rotating Venus! It had been proposed in the mid-1960s that solar heat absorbed in the clouds might be transported in a single cell away and *down* from the portion of Venus directly under the sun. It was hoped that such un-Hadley-like motions might produce the sizzling ground temperatures without having to rely upon the then-controversial greenhouse effect. Downward heat convection is no longer required to explain the surface heating, which is fortunate since it is difficult to imagine warm air sinking. Furthermore, Venus's atmosphere is so thick that it cannot respond to the daily radiation of the sun, even though "daily" for Venus means several Earth months. Rather, Venus's dense atmosphere senses the sun's warmth as a cover surrounding the planet, more intense at the equator than at the poles. Theorists began to expect that there might be traditional Hadley-cell winds, with greenhouse-warmed air rising in the Venusian tropics. But after Pioneer Venus, the Hadley model remains a plausible hypothesis in search of confirmation. Temperatures and winds deep in Venus's atmosphere vary more than had been expected prior to Pioneer. Complex eddy motions, or "weather fronts," may be at least as important as an average Hadley circulation for carrying equatorial warmth to the poles.

The visible clouds constitute only the tiniest, most tenuous part of Venus's upper atmosphere. Their motions may be influenced by the underlying winds, but they do not exhibit the simple motions expected for the top of the Hadley cell. We should not be surprised, however, since the winds in the Earth's stratosphere and above differ greatly from those we measure at the ground.

The first indication of cloud motions on Venus came 50 years ago from photographs taken through filters that admit only ultraviolet light. Unlike normal photographs of the blank cloud deck, ultraviolet pictures show changing patterns of dark patches on Venus. Not until the 1960s was it finally recognized that the patches move around Venus once every 4 Earth days, 60 times faster than the planet itself turns, or at a speed of 110 meters per second (250 miles per hour).

Are there really such strong winds in the stratosphere of Venus? Common sense may suggest that the rushing cloud patterns reveal how fast the winds must be blowing. But think a bit about familiar motions, real and apparent. Tap a taut rope and a wave rushes along the rope; yet the rope doesn't move horizontally. As you bob up and down in ocean waves, look at the water: while the waves sweep relentlessly to shore, the water stays pretty much where it is. Or consider a cloud suspended over a mountaintop while a stiff breeze blows right through it; obviously the cloud is not a wind marker like a balloon, but marks the position of a so-called standing wave as the air flows over the topographic obstacle. So do the Venusian clouds blow with the winds or do they mark illusory waves? Or both?

Mike Belton studied Venusian cloud motions from the beautiful ultraviolet closeup photographs taken by Mariner 10. The largest global features are horizontal, Y-shaped dark patches very reminiscent of the ultraviolet features photographed from Earth. He concluded that the moving patches represent wave propagation, not the actual motion of Venus's air. Yet the waves apparently move relative to the air at only a few tens of meters per second, much more slowly than the air itself moves around Venus. Just as the motion of a lantern carried by a man walking on a flatcar is due more to the train's speed than to the man's gait, so most of the 4-day apparent superrotation of Venus's atmosphere is real wind motion after all, and not chiefly the illusory motions of Belton's waves. Analysis of Pioneer Venus data shows that, at the level of the clouds, there is a Hadley circulation superimposed on the 4-day superrotation. The combined motions seem to form a polar vortex.

Pictures from Mariner 10 and Pioneer Venus show long, narrow belts near the equator of Venus that gradually shift southward. Where the sun is roughly overhead, there are cellular cloud structures, some with dark interiors, others bright. Near this subsolar region there often are long, tilted streaks that resemble bow waves around a ship, or shock waves generated by an obstacle in supersonic flow. Belton found terrestrial analogs for all these features. Some were recognized on Earth only since satellites began photographing our own planet regularly. The cells, he believes, are caused by convection under conditions analogous to shallow maritime inversion layers in the Atlantic and Pacific, which generate similar cloud patterns. The convection disrupts the rapidly flowing circulation of air around Venus, generating the bow waves. It will take a long time to understand the full variety of waves and eddies that occur on Venus. While the Hadley-cell perspective is useful and has some validity, the true behavior of Venus's atmosphere seems more complex the more we study it.

The major puzzle that I referred to a few paragraphs back still remains: What are the dark ultraviolet spots that circle around Venus every 4 days? By analogy with weather satellite pictures of Earth, they might seem to be holes in the clouds. But the sulfuric acid clouds have properties more like a smoggy haze layer than discrete clouds, and spectroscopic studies show that the spots are at the same altitude as the haze, imbedded within it. Rather small temperature variations cause terrestrial water vapor to condense or volatilize readily when the humidity is high. But sulfuric acid is very nonvolatile, so what makes the propagating waves on Venus visible? There would seem to be something in the clouds the thermodynamic behavior of which enables minor changes in temperature to change it from a clear gas to an opaque cloud. Among the candidate compounds proposed have been pure sulfur, bromine dissolved in hydrobromic acid, and—more recently—sulfur dioxide, chlorine gas, or nitrosylsulfuric acid. Nobody finds any suggestion entirely convincing, and it almost seems simpler to imagine that some cosmic artist is dabbing ultraviolet paint on some of the clouds of Venus!

Venus, the Earth, and Mars are all similar planets, although Mars is somewhat smaller. They may all have been made of roughly the same chemical mix of materials. All have probably formed an iron core. And all three planets absorb roughly the same amount of solar energy; Venus, though closer to the sun, reflects away a greater percentage of sunlight while Mars, which is farther away, absorbs more. We are challenged, then, to understand why the atmosphere of Venus is 100 times as dense as our own and why that of Mars is more than 100 times less dense than ours.

Have the atmospheres of the Earth, Venus, and other planets been different from the beginning? Or have they evolved radically since planetary formation? Clues that may provide a window on the past are beginning to emerge and challenge some conventional wisdom, thanks in part to the Viking Mars-landers and the Pioneer Venus descent probes. Atmospheres are composed of "volatiles": compounds that exist as liquids or gases at the temperatures of planetary surfaces. Volatiles are mobile, diffusing through the interiors of planets and floating up toward space. Some elements are chemically reactive and may exist in either volatile or nonvolatile form (carbon and oxygen form both carbon-dioxide gas and carbonate rocks). Others—the so-called noble or "rare" gases—are always gaseous except at extremely low temperatures and are completely inert. By measuring volatile abundances on planets today, and by studying the various chemical, physical, and geological processes that affect each compound differently, scientists can try to read backward to the beginning.

A classic question about our atmosphere has been whether it is of "primary" or "secondary" origin. As planets coagulated from the solar nebula, they may have gravitationally held onto a portion of the nebular gases. Although lighter gases such as hydrogen and helium would have escaped rapidly to space, and others would have been chemically bound up in underground reservoirs, the remaining gases would be considered a "primary" atmosphere. On the other hand, planets may have accumulated into large bodies long after nebular gases had dissipated. Or they may have formed at such high temperatures that their primary atmospheres evaporated into space. In such cases, their present atmospheres would have to be of "secondary" origin, having leaked out later from their interiors.

From the perspective of this simple dichotomy, there is little doubt that the Earth's atmosphere is chiefly of secondary origin. The composition of the sun, which must have "glommed onto" a representative group of nebular gases, differs from that of the fresh exhalations from volcanoes. Just the amount of neon in our air proves that very little air could be of solarlike composition. Neon, a noble gas, is chemically inert, cannot be stored within the Earth, and is too heavy to escape into space, so the Earth's total endowment of neon is in the air. If we were to add to the existing neon all other gases in their solar proportions, such a solar-type atmosphere would be less than 1 percent of our present atmosphere. There is simply far too little neon around for our air to be primordial.

Of course, atmospheric gases, like planets as a whole, must originally have come from the solar nebula. Differences between primordial and planetary compositions must reflect fractionations that occurred during planetary formation. By studying the differences we can learn more about those formative processes. One way planets might have become fractionated is if they accumulated strictly from solid bodies. Then their potential complement of volatiles would have been restricted to (a) those that can be formed from elements that were chemically bound up in the solid minerals (plus ices, for planets formed in colder parts of the solar system), and (b) gases with an affinity for being physically attached to the surfaces of solid grains by *adsorption*. When meteorites were the only extraterrestrial matter with a composition we could measure, it was thought that the relative proportions of their noble-gas constituents (when corrected for any solar component) were representative of early solid bodies (planetesimals) from which the planets accumulated. Hence, meteoritic compositions were termed "planetary." The Earth's noble gases are roughly "planetary," and it was predicted that Mars and Venus would have comparable endowments of volatiles bearing the "planetary" signature. Actually it was thought that Venus would be a bit volatile-poor, because it formed nearer the sun, and Mars a bit volatile-rich because it formed in a cooler place.

Once again, planetary spacecraft have proven that Nature is fickle in adhering to the theories of mortal minds. First, it was found that compared with Earth, Mars has only one-tenth as many atoms of argon-36 per atom of argon-40 (the number is the atomic weight). Now, ^{40}Ar is a radioactive-decay product, thought to be produced in similar amounts within all rocky planets. Since the chemical and physical behaviors of the two argon isotopes are nearly identical, Mars's relative lack of ^{36}Ar is a reliable indication that it was underendowed with primordial inert volatiles—the opposite of what had been expected.

Then Pioneer instruments measured a hundred times as much ^{36}Ar on Venus compared with the Earth! How could a world expected to be volatile-poor have accumulated so much ^{36}Ar? Comets and asteroids are volatile-rich and so an early bombardment of Venus would seem to be an answer. Such bodies have a lot of ices and other volatile-making elements such as carbon. Yet Venus has hardly any more carbon than the Earth (of course, in the case of broiling-hot Venus, it's mostly in the atmosphere, while on Earth, it's nearly all locked up in carbonate rocks). Also comets and asteroids strike Mars and the Earth about as often as they strike Venus, so unless Venus were struck by one large ^{36}Ar-rich body—a statistical fluke—we must look elsewhere to understand the Pioneer data.

Other noble gases such as krypton are not present on Venus in the preconceived "planetary" proportions. Indeed, George Wetherill has suggested that the planetesimals that formed Venus might have been caught in a brisk solar wind that implanted their surfaces with a large solar component of volatiles, and that these volatiles are now part of its atmosphere. If so, why did the planetesimals that accumulated to form the Earth, Mars, and the meteorite parent bodies fail to get a solar component, too? Either we don't understand how noble gases and other volatiles were injected into solid planetesimals, for wide variations from place to place in the solar system are not thought to have existed, or the planetesimals in one region were somehow degassed while those in another were not. As we unravel these chemical mysteries, we will learn a lot about how the Earth and other planets were made.

While the noble gases are important in scientific detective work, they have little practical importance for sustaining life. More important gases—nitrogen, oxygen, carbon dioxide, methane, and ammonia—were incorporated in the planets mostly by being chemically bound into solids. Evidently planetary formative processes were not nearly so selective in distributing such compounds among Venus, Earth, and Mars as they were in allocating the rare gases. It is thought that Venus and the Earth have similar inventories of these chemically reactive volatiles. Viking data on isotopes of nitrogen and oxygen prove that Mars was originally endowed with an adequate quantity of compounds to form a substantial atmosphere. Why, then, are the atmospheres of the three planets so different? The answer lies in chemical interactions, which, in combination with physical, geological, and biological processes, have greatly exaggerated the minor differences among the three planets.

It seems that Venus was just enough warmer than Earth to cause an irreversible sequence of events that boiled away its oceans and led to the present inferno. On Earth, water remained stable and provided a rich environment for the origin of life; life, once established, changed the chemistry of the air and water to the conditions familiar today. Mars is a bit colder and smaller than Earth, so some of its volatiles have been frozen or trapped into underground and polar deposits, while others have leaked off the top of the atmosphere into space because of the weak Martian gravity.

Atmospheric chemicals are prone to react with each other and with the ground until everything is in chemical balance and no further reactions occur. But the Earth's atmosphere is in a state of constant chemical activity, never reaching equilibrium, as new rocks are rapidly produced by volcanic and mountain-building processes; the winds and rains wash away the chemically broken-down or weathered rock, always exposing fresh rocks to the air. Radar images of mountain ridges and canyons suggest there has been great geological activity on Venus, just as is certainly true of some Martian provinces. But there are no oceans on Mars or Venus in which to deposit sediments. There are not even rains or streams to wash away soils, though possibly there were on Mars in times past. A corrosive drizzle may fall through the sulfuric acid clouds of Venus, but it never reaches the ground. Perhaps winds on Mars and Venus are sufficient to scour away soils continually, so that their atmospheres eventually may approach chemical equilibrium with rocks throughout their crustal layers. If not, reactions involving just the surface layers of the planets may have had only a superficial effect on the atmospheric chemistry. Even the great heat of Venus, which is conducive to rapid chemical reactions, would not have been able to overcome an absence of erosive processes, thus leaving the atmosphere chemically isolated from the bulk of the planet.

Planetary atmospheres can depart from chemical balance because of processes that occur in their uppermost regions. If the temperatures are high enough, or reactions provide sufficient "kicks," atoms may be accelerated beyond escape velocity. Over the eons great quantities of the lighter gases may evaporate into space as they are replaced by upward diffusion from below. A particularly potent influence on upper atmospheric gases is ultraviolet solar radiation, which is sufficiently energetic to tear a water molecule into its constituent parts of hydrogen and oxygen. Hydrogen, the lightest gas, evaporates rapidly. Sunlight can also photodissociate other gases. For instance, carbon dioxide is converted to carbon monoxide plus oxygen, which are then available for further reactions. The steady-state composition of an atmosphere depends on the rates of disequilibrating processes relative to reaction rates and relative to rates at which winds mix the atmospheric gases.

A major disequilibrating process is life itself, which has had a profound effect on our own air. Long ago the Earth's atmosphere was much richer in carbon dioxide. It also contained traces of hydrogen and hydrogen-bearing, or "reduced," gases such as ammonia (NH_3), thought to be necessary for the origin of life. The Earth's atmosphere originally lacked oxygen, including the rich form we call ozone, which blocks out ultraviolet sunlight. Unshielded solar radiation may have provided the energy necessary for synthesis of life in the unoxidized oceans. Photodissociation of water in the early stratosphere produced some oxygen, but it was consumed by fresh oxygen-poor gases released into the air from volcanoes. Eventually, about 2 billion years ago, lifeforms evolved, such as blue-green algae and more complex varieties that produced oxygen by photosynthesis of carbon dioxide, and oxygen finally began to accumulate in the Earth's atmosphere. Oxygen is now maintained in the air by the burial in ocean floor sediments of dead phytoplankton which, if they were not buried, would be oxidized and thus consume the oxygen they had originally photosynthesized near the top of the ocean waters.

Oxygen, carbon dioxide, and water, being in chemical disequilibrium with rocks, help to weather rocks into soils that are washed into the sea. Some of these in turn are used by ocean life to form shells which, along with chemically precipitated carbonates, accumulate into great deposits of limestone and other carbonate rocks on the ocean floors. At continental margins the sea-floor-spreading treadmill carries the rocks down into the Earth's hot mantle (see Chapter 12). From the descending plates, molten lavas rise to the surface, disgorging great quantities of carbon dioxide and other gases which, having reached chemical equilibrium with the interior of the Earth, are out of equilibrium with the surface and once again begin to eat away at the rocks.

The quantity of carbon contained in carbonate rocks in the Earth's crust is roughly the same as that contained in the carbon dioxide in Venus's massive atmosphere. Venus lacks water and may lack plate tectonics, which may mean that all the carbon dioxide that has leaked out of its interior remains trapped in the atmosphere because it has no way of participating in a geochemical cycle with the Venusian crust. Alternatively, at the broiling temperatures at the surface of Venus, 100 Earth atmospheres of carbon dioxide may be the equilibrium value (we can't know for lack of good evidence about the minerals that compose Venus's rocks). But the high temperatures on Venus are partly due to the vast atmosphere of CO_2, so the question as to how it ever got that way to begin with remains.

The answer seems to lie in the evolution of water on Venus. The one chance for Pioneer Venus scientists to study water in the bone-dry atmosphere of Venus came from data recorded during the "accident" when that droplet of sulfuric acid unexpectedly blocked the inlet to an instrument. Three years later, John Hoffman and Thomas Donahue, of the universities of Texas and Michigan, respectively, suddenly thought of checking the data for deuterium—a heavy form of hydrogen. They found it enriched by *100 times* compared with deuterium in the Earth's water! Evidently modern Venusian water is a "heavy" residue of an oceanic volume of normal water, most of which must have escaped to space long ago.

So Venus received its fair share of water initially and thus originally exhaled a moist, carbon dioxide-rich atmosphere like that of the primitive Earth. But evidently Venus is just near enough to the sun, hence sufficiently warmer, that the water vapor created a runaway greenhouse effect. Its infrared absorptions helped to warm the surface, which vaporized even more water, which trapped still further heat. Soon the steamy atmosphere, containing the ocean's worth of vaporized water, approached current temperatures. Some of the water vapor was lost through reactions with carbon monoxide and hydrogen sulfide, the latter a gas that smells like rotten eggs. Ultraviolet sunlight broke down the remaining water vapor into hydrogen, the normal "light" form of which immediately escaped, and oxygen. The fate of the oxygen is uncertain; possibly it escaped as well, or more likely it oxidized crustal rocks or lavas. Anyway, once Venus got this hot, there was no way for it to cool again.

James Pollack has speculated that the runaway greenhouse may not have happened on Venus until quite recently in its history. Astrophysical theories concerning the evolution of stars suggest our sun may have been dimmer in the past and may become more luminous in the future. It is a thought-provoking notion that Venus may once have had rivers and oceans, a moderate climate, and possibly even life. Then the sun grew brighter, and within several hundred million years Venus became the hellish place it is today. Could Venus be a prefigurement of the future evolution of our own benevolent world?

The story for Mars is very different and somewhat uncertain. Both water and carbon dioxide are stored in the polar ice caps. Quantities greatly exceeding the amounts now present in the tenuous Martian atmosphere are trapped and frozen in the upper layers of the Martian ground. Some volatiles exuded into the atmosphere by the volcanoes, or by sublimation and melting of ice, were photodissociated by the sun. Certainly much water vapor has been destroyed this way, with hydrogen evaporating into space. As with Venus, the fate of the oxygen is uncertain; it may also have been lost at the top of the atmosphere or it may have helped oxidize the iron in Martian soils, producing the rusty color of the planet. Other volatiles have been lost by energetic photochemical reactions high in the Martian atmosphere or have reacted to form rocks. Probably much of the volatile inventory of Mars has either not yet been outgassed or has become trapped again in underground reservoirs, leaving the Martian air thin.

Yet the widespread channels on Mars testify to an earlier epoch when the atmosphere apparently was different and waters may have flowed. If the crater erosion and channelization occurred early in the history of Mars, before a thick atmosphere evaporated into space or became permanently imbedded within the ground, we might account for rivers as a once-only event in Martian history. Less likely but more interestingly, a wet period may have occurred rather recently, during a temporarily warmer climate; the ice thawed and the ground degassed, generating an Earth-like atmosphere for a time, before cooling temperatures froze the ice again to await another future warming. We will look at these possibilities again in Chapter 12.

Although Venus seems to be pretty much trapped in its present hot, acidic state, the atmospheres of both Earth and Mars will continue to evolve. With the arrival of technological man on the scene, that evolution may be very rapid, indeed, for until human beings evolved, the biosphere and atmosphere of the Earth lived in a harmonious, symbiotic chemical relationship. Now we have introduced great disequilibrating changes that may, or may not, drastically alter the character of our life-sustaining envelope. If some theorists are correct, the Martian atmosphere may be even more susceptible to human influence—for good or ill. Before anyone tries to start runaway atmospheric evolution on Mars, we should all consider the merits of designating Mars an international wilderness area, to preserve for eternity the history imprinted in its rocks of a more Earth-like time, when Martian life might have thrived.

8

The Moon: What Did We Learn from Apollo?

The termination of the Apollo flight program after Apollo 17 leaves the scientific tasks undertaken by Apollo substantially unfinished.
—*Report of the Lunar Science Institute, 1972*

A mansion stands on the marshy coastal plains of eastern Texas, where the cultures of the American West and South mingle. A proud relic of bygone days, its balustraded patio overlooks Clear Lake, an estuary of Galveston Bay. Beyond formal gardens, now in disrepair, lies a tree-encircled private pond of several acres. A chandelier descends from the double-storied ceiling of the central hall and a broad, double staircase sweeps down toward the intricately carved fireplace. In these comfortable quarters, once the heart of the ranch of "Diamond Jim" West, lunar scientists from around the world gather to meet and talk and try to fathom the results of the greatest scientific enterprise in the history of mankind.

Emerging from an evening seminar in the Lunar and Planetary Institute (formerly the Lunar Science Institute), few conferees notice the yellow orb rising between the branches, glowing dully through the sultry Houston smog. For them, the meaning of the Apollo Project and the conquest of Earth's satellite lies in the collection of exotic moon rocks, stored a mile away in one of the many buildings that compose the Johnson Space Center. As the years of lunar exploration recede into the past, the measurements and analyses continue, fostered and guided by the government-supported, university-controlled institute which occupies the restored West mansion.

Nobody's going to the moon anymore. For most Americans there are dim memories of Mission Control, a golf game amid the craters, and parachutes splashing in the Pacific. Maybe they recall questions of how old the moon was, whether it was hot or cold, and whether the meteor craters were actually volcanoes. And how did the moon form, and why? Perhaps people now take for granted that the answers were learned, although the news never made headlines. The news media soon turned to Cambodia and Watergate, then energy and inflation, leaving Apollo scientists to study their moon rocks outside the limelight in which they had briefly been.

Few taxpayers realize that they are still spending several million dollars a year to learn about the nature and origin of the moon. But that wealth is but a dribble compared to the flood of expenditures in the 1960s that brought the moon rocks back. As the dribble dwindles, and one by one the lunar researchers return to studying meteorites and ocean basalts, the time has come to ask, "What have we learned?"

Did we find those pristine rocks from the earliest years of the solar system? No, we didn't. Is the moon hot or cold? It's warm. Were the craters formed by impacts or volcanism? Pre-Apollo opinions of the majority favoring impact haven't been changed, nor those of a few dissenters. How was the moon formed? Well, we still don't know. Then was the Apollo program uproarious, spectacular, and expensive, but yielding few results, and of no ultimate benefit to taxpayers? I think such a conclusion very wrong. For the scientists who congregate at the Lunar and Planetary Institute, the exploration of the moon has been a terrific success. The benefits of Apollo are spreading throughout the physical and geological sciences. That we didn't learn immediately why the Earth has a moon is but a trifling detail to satisfied researchers. As for taxpayers, we must all stand back and examine the Apollo project and its relationship to science in fair perspective.

Why did we go to the moon? It is a simple question, but the answers are complex, involving politics, psychology, economics, and even foreign policy. It is a cliché, but well worth repeating, that the reason we went to the moon was certainly *not* to do scientific studies of the origin of the moon. Science, which never has been more than a tiny facet of human culture, would never by itself have rated the expenditure of tens of billions of dollars. At best, science occasionally rides piggyback on larger endeavors; such was true of Apollo.

What motivated Congress to vote the funds to go to the moon? That is the crucial question, because a necessary and probably sufficient requirement for going to the moon was the allocation of funds. Here is why congressmen said they voted the funds: "The United States has not embarked upon its formidable program of space exploration in order to make or perpetuate a gigantic astronautic boondoggle. There are good reasons for this program. But, in essence, they all boil down to the fact that the program is expected to produce a number of highly valuable payoffs. It not only is expected to do so, it is doing so right now. . . . Those already showing up . . . include the most urgent and precious of all commodities—national security."*

Whether national security was the prime mover or not, it is most unlikely that science was. Once the money was allocated, however, one might have expected a scientific orientation from the National Aeronautics and Space Administration, which ad-

* From *The Practical Values of Space Exploration,* Report of the Committee on Science and Astronautics, U.S. House of Representatives, Rept. No. 2091, 86th Congress, 2d Session, 1960.

ministered the program. But even within the one small branch of NASA that was responsible for science, other motives predominated in the early years of Apollo. In 1963 the then-director of the NASA Office of Space Science wrote: "While science plays an important role in lunar exploration, it was never intended to be the primary objective of that project. The impetus of the lunar program is derived from its place in the long-range U.S. program for exploration of the solar system. The heart of that program is man in space, the extension of man's control over his physical environment. The science and technology of space flight are ancillary developments which support the main thrust of manned exploration, while at the same time they bring valuable returns to our economy and our culture. . . . Thus the pace of the program must be set not by the measured patterns of scientific research, but by the urgencies of the response to national challenge."

Even today, with NASA's budget far less than what it once was, "man in space" remains the overall purpose of the space program. There is a powerful emotional, even spiritual, impetus behind man in space. It is captured in the novels of Arthur C. Clarke, and it springs from the same motivations, endemic to Western culture, that have always impelled us to conquer the wilderness and physically transform it. Or one could be more cynical and suggest that NASA is like any other federal agency. Its power depends on its cash flow, and expensive hardware needed to shoot people into space is most important, not inexpensive science.

NASA administrators, and some scientists too, offer many rationalizations as to why the best way to study the solar system is to send human beings out there. But the arguments ring hollow, except to the true believer. Nowadays, space scientists watch the Space Shuttle project gobble up the bulk of NASA's allocations from Congress while a once-in-a-lifetime opportunity to study Halley's Comet is bypassed. These scientists lobby, ineffectually, for a reorientation of NASA's priorities that almost certainly will never come.

Other scientists, perhaps more practical and certainly more attuned to political realities than their more idealistic colleagues, support the Shuttle and the man-in-space activities. They believe that the little money left over for science from such enormous technological projects, and the occasional piggyback opportunities, will provide an adequate base for scientific research. If man-in-space projects were threatened, NASA might collapse and space scientists would be left holding an empty bag. Scientists may be left with nothing anyway, as ever-deepening budget cuts slice into NASA's programs. NASA will have built a "truck" (the Shuttle) and a "warehouse" (the proposed Space Station) but will have lost its purpose (exploration of the solar system) and will be unable to provide a "cargo" (spacecraft and instruments). No doubt the military will be happy to use and occupy the products of NASA's endeavors.

Our headlong rush to the moon was not a scientific research program and the astronauts (except for one) were not scientists. Once Neil Armstrong had taken his "one small step for a man" and the Americans had clearly beaten the Russians, public interest in moon landings nosedived and the series of subsequent Apollo missions was cut short. Now NASA can't even get the money to support one unmanned, polar-orbiting lunar satellite, designed to follow up on Apollo discoveries.

But if science wasn't important for Apollo, Apollo nonetheless revolutionized solar system science. The exhilaration of lunar scientists during the heyday of Apollo can never be recaptured. Theories debated for centuries were being resolved every week and hundreds of new mysteries were being revealed. Seismographs on the lunar surface were telemetering back totally unexpected reverberations deep below ground level. Lunar soils seemed unaccountably to have ages hundreds of millions of years older than the rocks from which they were derived by meteoritic erosion. The amount of data returned from the moon, both in sample bags and radioed directly back, during the 3½ short years from Apollo 11 to Apollo 17 was truly staggering for scientists accustomed to waiting months for one small meteorite to fall from the skies.

Lunar science has now matured. The initial production-line measurements have all been done. The thoughtful, measured pace of scientific research has returned. There has been time for contemplation and for ingenuity to invent new techniques to unravel the history locked up in the lunar sample collection. The bright, imaginative physicists and chemists lured into the Apollo project now have their sophisticated instruments built and working, and reliable measurements are being reported and confirmed by other laboratories. Although the excitement of sample acquisition is gone, the serious process of experiment and synthesis is well underway, and new insights to the origin of the Earth and the moon are being reported at every meeting convened at the Lunar and Planetary Institute. Recently, however, shortsighted budget-cutters have considered curtailing most of that research and padlocking the doors of the Institute. Money would even be withheld from caring for our invaluable collection of moon rocks. Thus the entire creative research effort might be cut off in midstream. At times it is difficult indeed to comprehend the logic of Washington decision-makers.

The pre-Space Age picture of the moon was very incomplete. Its interior was a complete mystery. We could only observe radiation reflected by, or emitted from, the front side of the moon. The few clues we had about the moon's composition were indirect and inconclusive. We had pictures chiefly of its craters and plains. As augmented by intensified telescopic observation in the 1960s and the exploratory Ranger, Orbiter, and Surveyor spacecraft, our hazy pre-Apollo hypotheses for lunar evolution mainly concerned its large-scale surface geology.

The details in the moon's story have always been uncertain. Half a billion years after the moon's formation 4.6 billion years ago there was a torrential bombardment of the moon. Catastrophic impacts excavated great basins, throwing debris far across the moon and scouring preexisting craters. Subsequently the basins were flooded by lava flows, which cooled into overlapping layers of dark basaltic rocks. In the mid-1960s William Hartmann, then at the University of Arizona, estimated the period of volcanism occurred about 3.6 billion years ago, or 1 billion years after the moon had formed. He was right; we now know it started at least 3.9 billion years ago and lasted at least 700 million years. There was occasional faulting of the lunar crust, but there were never even the beginnings of Earth-like crustal plate motions. During the last 2 or 3 billion years the moon has been struck and cratered by countless small bodies, and a few rather large ones, but otherwise its surface has been geologically dead.

These pre-Apollo ideas have generally survived the more extensive photography by the orbiting Command Module pilots and the close-up examination of several locales by astronauts on the surface of the moon. Of course, intelligent and imaginative scientists are adept at incorporating new data into their preconceived hypotheses; it is harder to be sure of the truth. Also, with so many data to assimilate, scientists are often first inclined to accept uncritically the results of others outside their own narrow experimental or theoretical specialties. As the inconsistencies are gradually ironed out, some accepted interpretations of lunar geological history are sure to change.

While our views of lunar geology have only been sharpened by Apollo, entirely new and fundamental sides of the moon's personality have been revealed. These have come mainly from studies of the chemicals (and their isotopes) and the minerals of which moon rocks are made. We now know what the rocks are made of, and we think we know their origins. Since all moon rocks seem once to have been molten, we can deduce something of the thermal history of the moon. Some rocks must have been derived from others at great depth in the moon; therefore, just as the shape of a jigsaw puzzle piece reveals the shapes of its neighbors, the compositions of surface rocks provide clues about the lunar interior. The nature of the interior is further revealed by the way it affects the electrical and magnetic fields of the solar wind and in the way it propagates heat and seismic waves. The ages of moon rocks have been measured. Each rock records several ages, some reflecting the origin of the moon, others the gross separation of the moon into layers, and still others the subsequent volcanic and impact episodes that affected the different lunar provinces.

All rocks on Earth are composed entirely of minerals, each of which is a substance having a specific crystalline structure and chemical composition. We now know that moon rocks are made from some of the same minerals familiar in terrestrial and meteoritic rocks. That was no surprise to geochemists, who had never expected to find green cheese or any other exotic substance, but had no previous idea about which of those minerals to expect and in what proportions. The way in which chemical elements form mineral compounds reveals chemical abundances of the precursor material and the subsequent temperatures and pressures to which it has been subjected.

A common pre-Apollo idea was that the moon was made of chondritic meteorites, which contain the same proportions of most chemicals as the sun itself. As I have described earlier, the chondritic meteorites are thought to be the unaltered and unmelted materials that condensed from the original nebula in at least one part of the young solar system. Had the moon formed cold from chondrites and not been melted since, moon rocks would be chondrites. They aren't. All the stable chemical elements occur in lunar rocks, if only in trace amounts of parts per billion. Compared with chondrites, lunar rocks grossly lack some elements by factors of more than 1,000, yet are enriched in some others by factors of 10 to 30. Either the moon was not made out of chondrites or the chemicals have gotten rearranged on the moon so that the surface rocks are not representative of the whole moon. Or both.

It might seem presumptuous to pick up rocks from the moon's surface and claim to know what the bulk of the moon is made of. But geochemists have long studied the Earth's rocks and have developed pretty good ideas about what kinds of processes enrich certain chemicals and deplete others. For instance, as hot gas of solar composition cools, the first minerals to condense when the gas "cools" to 1,500 K are calcium-rich compounds, followed by metallic iron, then the more common silicates, with water appearing relatively late (at the boiling point of water). So if the moon contained just those minerals that condense above a certain temperature, but none of the lower temperature compounds, we could conclude that it condensed from a hot part of the solar nebula and has not been modified since. No other known chemical process would select chemicals in just the same way. The moon actually is enriched in many of the high-temperature compounds and depleted in some of the more volatile ones, but that is by no means the whole story.

Gold and nickel are absent on the moon, despite these elements' moderately high condensation temperatures. Since gold and nickel have a strong affinity for becoming alloyed with iron, it was immediately suggested that surface rocks are depleted of these minerals because they sank into a metallic core when the moon was molten. Yet we know from the lunar gravity field that a massive metallic core simply does not exist in the moon. It seems that the separation of iron-related elements from the moon occurred *before* the moon was made. I will return to just how the moon was made later, but for now we can accept that the moon was never chondritic but is enriched in high-temperature elements and depleted of volatile and iron-related elements.

There are still further ways in which the chemistry of moon rocks distinguishes them from meteorites, from terrestrial rocks, or even from each other. Most of these characteristics are very well known to geochemists familiar with rocks that have once been molten. For example, there is a group of chemicals known as the rare-earth elements which, because of their atomic sizes and poor ability to unite chemically with other elements, are poorly incorporated in mineral crystals. Thus when rocks are heated only a certain amount, a portion of the material melts into a liquid that is rich in rare-earth elements. Being composed of many different chemicals and minerals, rocks do not have a single melting temperature; rather, an increasingly large portion of the less compatible constituents in a rock enter the melt as its temperature is raised over a range of hundreds of degrees. The rocks from the lunar maria are rich in such incompatible elements as the rare earths, strongly indicating that they are resolidified lavas, originally derived by the partial melting of deep interior rocks. Indeed, just as one can tell what meat was cooked by tasting the gravy, geochemists can infer the composition of parent rocks by analyzing the melts.

One rare-earth element, europium, acts differently from the

rest. It seems to have been greatly depleted from the basaltic lavas that flooded the lunar maria basins. The depletion is probably due to europium's strong affinity for feldspar, a mineral rich in aluminum and calcium. It happens that in contrast to the maria rocks the lunar highland rocks are rich in feldspar and have an excess of europium. When molten rocks are cooled, experiments show that feldspar crystals are the first to form and tend to float to the surface. Their predominance in highland rocks, even those excavated from great depths by cratering impacts, first suggested to lunar scientists that the europium-enriched highlands formed from the top of a cooling ocean of molten rock, or magma. Evidently the europium-depleted maria lavas came later from the partial remelting of rocks that had settled toward the bottom of the original lunar magma ocean. Thus the complementary character of europium and other trace elements in maria and highland rocks reflects a period ending about 200 million years after the formation of the moon, when at least the outer layers of the moon were substantially molten and a gross chemical separation occurred.

How deep was the magma ocean? And from what depths were the lavas subsequently derived 3.2 to 3.9 billion years ago? These questions are central to our understanding of the moon, and there are some tentative answers. Actually there is increasing evidence that there never was one great magma ocean, pure and simple. The composition of the lunar crust is not the same at all longitudes. And the upper regions of the moon may have been slushy rather than completely molten. Perhaps there were several smaller magma oceans, or large "lakes," rather than a global ocean. But whether the upper layers of the moon were completely molten or not, lunar scientists still find it convenient to approximate complex reality with the simpler concept of a magma ocean.

Velocities of seismic waves, caused by both tiny natural moonquakes and an occasional rocket or meteoroid striking the moon, indicate that the feldspar-rich rocks underlie both the highlands and the maria to a depth of about 60 kilometers. A magma ocean with an upper crust 60 kilometers thick must have been very deep indeed! That fact is confirmed by the presence of great mass concentrations, or "mascons," that pull on spacecraft orbiting the moon. The mascons seem to be the thick, dense, solidified lava lakes that fill the circular mare basins. They are so heavy that they could not be supported by the deformable, partially molten rock from which the lavas were derived. Were the partly molten layers close to the surface, the mascons would have sunk into the moon and would not remain today. The mascons must have been supported by a cold, rigid shell or crust well over 100 kilometers thick, so the lavas came from at least 100 kilometers down. But they cannot have come from much deeper than 400 kilometers, where the pressure is about 20,000 Earth atmospheres. The only kind of rock at such enormous pressures that could produce melts of lunar basaltic composition would be rich in garnet gems and have a density far greater than is permitted by measurements of the lunar gravity field. So the lavas came from depths of 100 to 400 kilometers, and 400 kilometers evidently marks the bottom of the hypothesized magma ocean.

The lunar lava rocks differ somewhat in composition. Those found at the Apollo 11 and 17 sites contain a lot of titanium, while those at the Apollo 12 and 15 sites are titanium-poor. Since the former are ½ billion years older than the latter, geochemists have looked for ways in which the source region might have changed with time. Perhaps the temperatures of the parent rocks changed, producing melts of different composition. Alternatively, different parent rocks—perhaps at different depths—might have melted at different times.

There is a plausible scenario for the evolution of the lunar lavas that is consistent with what we imagine to be the thermal evolution of the moon. Since the moon is so small, compared with the larger rocky planets, it cannot retain its interior heat for long. So it cooled, especially at the surface. From the presence of the mascons we know that the upper 100 kilometers of the magma ocean had cooled within the first 700 million years. Presumably the outer layers continued to cool, and what heat was left migrated inward, where it was insulated by the thick lunar crust. The heat pulse eventually reached the core of the moon, which is evidently still quite warm today. The evidence for that is that a particular type of seismic wave that cannot move through liquids does not pass through the core of the moon; so the core is at least partially molten and well above 1,300 K.

The effect of the heat pulse on the melting rocks is the purview of experimental petrologists. These are scientists who subject rocks to high pressures and temperatures and study resulting changes in the minerals of the rocks, including the compositions of melts. Based on their experiments, they have developed models (theoretical scenarios) for the cooling of the early magma ocean that would explain the sequence of lava compositions. They believe that the rocks that could sweat out the high titanium lavas would be sandwiched at intermediate depths in the moon—below the feldspar-rich crust but above the kind of rocks that would give rise to low-titanium basalts. So as the heat pulse slowly propagated downward in the early moon, there were first high-titanium flows, then low-titanium ones, consistent with the measured ages of the rocks. Finally, 2 or 3 billion years ago, no lava at all could penetrate the thickening lunar crust, all lunar volcanism ceased, and the lunar surface became inactive.

It is generally true of lunar science that the facts and possibilities are so numerous that no simple model is entirely adequate. That is true of the magma ocean concept. So, it is too with this scenario for the generation of lavas. Basalt titanium content has been inferred from telescopic measurements for regions never visited by the astronauts. The technique, described in Chapter 4, has been tested for regions visited by the astronauts and seems to work. Yet a crater-counting age-dating technique suggests that some of the apparently most titanium-rich regions of the moon are not old but rather are among the youngest, younger than any rocks brought back by the astronauts. So either the telescopic measurements are wrong, the age-dating technique is wrong, or the models for generating lavas at great depth in the moon are wrong.

In order to help resolve these questions, the Lunar Science Institute (as it was then known) convened a conference in November 1975 entitled "Origins of Mare Basalts and Their Implications for Lunar Evolution." The ingenuity of the participants gave birth to numerous alternative models to explain mare basalts. Some theorists proposed that the basalts were derived from unmodified, primitive lunar material at great depths in the moon, well below the early magma ocean. Others found ways to produce lavas from very shallow depths in the moon. Even more exotic mechanisms were invoked to explain the local concentration of highly radioactive rocks called KREEP (rocks that are rich in potassium [*K*], *R*are *E*arth *E*lements, and *P*hosphorus).

In the late 1970s, the Institute added "Planetary" to its name at the same time that it initiated a large project, involving nearly 100 scientists, to compare lunar volcanism wih terrestrial volcanism and with what we know of volcanism on Mercury, Mars, and even some asteroids. Geologists and astronomers who had never been interested in the moon before were enlisted in the hope that comparative studies, spurred by the exquisite and exhaustive work that had already been done on lunar volcanism, would lead to new insights about volcanism on all the planets, including the Earth. This fundamental stage of planetary evolution occurred early in the lives of such small bodies as some asteroids and even the moon, but it is at its peak right now on our own planet. Because of Earth's melting and chemical separation processes, analogous to those that occurred on the moon, we have available the rich concentrations of minerals called ores that have made possible much of human technology.

The basaltic volcanism project resulted in the publication in 1981 of a huge, definitive book, *Basaltic Volcanism on the Terrestrial Planets* (Pergamon Press). The creative interactions of scientists from many disciplines led to such new ideas as the realization that some unusual meteorites may, in fact, be volcanic rocks from Mars! It is hard to say whether or not there will be many practical applications from our better understanding of volcanism that can help recoup the vast expenditures of Apollo. But the comparative planetological approach being fostered by the NASA Office of Space Science through the Lunar and Planetary Institute is clearly a first step in reaping the practical long-term benefits from learning about the moon. It remains for our country's uncertain political processes to determine whether or not there will be continued funds to support this important research.

Long, long ago, it is told, two Eskimo children were playing in their igloo during the eternal darkness of winter. The sister once chased her brother out into the darkness and they ran across the ground carrying their torches above them. Suddenly they were lifted into the sky, where they chase each other to this day. The girl is the sun. Her brother's torch is dimmer; he is the moon.

Modern science has had its own mythologies about the origin of the moon, some as quaint as the Eskimo legend. Some have said that the moon was wandering through the heavens from distant places when it came near Earth. The Earth reached out and captured it, forcing it always to face its captor. Others say the young Earth spun faster and faster as it formed until a huge chunk was flung out from the Pacific Ocean basin. As Eve was created from Adam's rib, so the moon was born of the Earth. Still others have imagined a twin birth in the Earth's orbit that yielded a double planet from the very beginning.

The wealth of data from Apollo has enabled the natural philosophers to embroider their simple tales to the point that they are stunning creations of human imagination. But as they all suffer from incompatibility with the laws of physics, or with the trace element abundances, or with other immutable facts, we can be sure that the full truth has yet to be perceived. Scientists are optimistic, however, that from all the new data, combined with further study of other celestial examples, such as moonless Venus and moon-laden Jupiter, we may soon learn for sure how it is our planet has a companion. And perhaps, even now, we can begin to see the glimmerings of the final answer.

I have already recounted how the moon rocks have been depleted of such metals as nickel and gold, which have a strong affinity for metallic iron and which would accompany iron into the core of a molten planet. In that sense the gross mineralogy of the moon resembles the crust of the Earth. Perhaps the moon was yanked from the Earth, with its iron-related elements already sunk into the core of the Earth. However, physicists think there are insuperable difficulties in getting a moon to escape as a whole body from the Earth's crust and evolve to its present orbit, leaving the earth spinning at its present 24-hour rate. In addition, there are great differences between the Earth's mantle and the moon in terms of the abundances of such volatile trace elements as bismuth, sodium, and gold. These differences would be difficult to explain by known chemical processes.

It seems that the moon must have been made out of materials somewhat different from those that composed the primordial Earth. Some theorists even proposed that the moon formed near the planet Mercury from minerals that condensed first from the solar nebula at very high temperatures. That would account for the enrichment of those minerals in the moon, but ad hoc arguments were required to explain why the moon wasn't also enriched in iron, which is also a very high-temperature condensate. Furthermore, the celestial billiard-ball wizardry necessary to get the moon accelerated away from Mercury and then decelerated into a nice circular orbit around the Earth was more than any celestial mechanician was willing to vouch for. Even capture of the moon from less exotic locations of the solar system is not easy to imagine. A body hurtling past the Earth would zoom away again, not go into orbit. The Earth's gravity would tug at the retreating moon but would slow it down by no more than the amount it had accelerated it on its approach. An object coasting down a hill gains enough speed to carry it up the next hill; if it gets trapped in the intermediate valley it is because of friction, which has no analogy in interplanetary space.

If simple capture is a highly improbable explanation for the moon, it would be totally impossible to believe that the dozens of other satellites in the solar system were also captured. It is far more reasonable to believe that moon formation is a natural part of planet formation. That aesthetic nicety, however, is about all the double-planet hypothesis has going for it. Physicists doubt that under normal circumstances any moon formed next to the Earth could evolve into an orbit of the present tilt. Furthermore, one might expect two bodies in the same orbit about the sun to be made of the same materials. Instead, the Earth is much denser than the moon, and I have already described other compositional differences between the two.

Currently theorists are attempting to amalgamate the better features of all three ideas, combined with new insights about the early solar system. It seems to me that an important element of any future model for the origin of the moon will be the intermixing of material throughout the early solar system. Moonlets, asteroids, comets, and other planetesimals left over from the later stages of the formation of Jupiter, Venus, and the other planets were flung about the solar system. The later stages of the formation of the Earth and the formation of the moon must have been greatly influenced by all these bodies. Because of the differing sizes and gravities of the Earth and the moon, it would not be surprising that different percentages of high- and low-temperature condensates were incorporated in the two.

The moon may even be largely composed of one or more bodies that came whizzing past the Earth in the earliest epochs. If a pass were close enough, which is quite probable, great tidal forces would have torn such a body apart. Some of the pieces would then have gone into orbit around the Earth while the rest flew off, never to return. If such a body had already melted, formed a core, and solidified, the core might have been among the portions that continued on their way, which would account for the lack of iron-related chemicals in the moon without requiring them to be in the Earth's core. Alternatively, one can imagine collisions occurring in near-Earth space between some protolunar material and intruders from elsewhere. Because stone fractures so easily, while metal does not, there could be a net separation of the metal from the stone that ultimately resulted in the coagulation of a metal-depleted moon.

There is also new evidence of a close genetic link between the moon and the Earth. It comes from a research tool developed after the Apollo program ended. Rocks are analyzed for the ratios of the two less common isotopes of oxygen to common ^{16}O. Isotopes differ in the properties of their nuclei, such as their atomic weights, but not in their chemistry. So ratios of the three isotopes should remain intact, even if major chemical processes enrich or deplete a certain material in oxygen. There are just the slightest separations of the isotopes caused by the slight differences in weight. Any such change in the ratio of ^{18}O to ^{16}O should be just twice the change in ^{17}O to ^{16}O, since the difference in weight is twice as great. Yet when University of Chicago geochemist Robert Clayton measured the isotope ratios for various rocks from the Earth, the moon, and the meteorites, he found a dozen different groupings that cannot be derived from each other by any chemical or physical process that could operate within a primordial solar nebula of uniform composition.

Oxygen isotopes were originally formed deep within stars, which subsequently exploded. Contrary to the canonical earlier view, it appears that heterogeneous star-stuff must not have been thoroughly mixed in the gaseous nebula. Presumably water vapor, carbon monoxide, and other oxygen-bearing gases had one isotopic composition. But there was another component, too, about 5 percent richer in ^{16}O, which got mixed together in different proportions with nebular condensates. Probably the ^{16}O-rich component consisted of presolar, interstellar solid grains that somehow never evaporated in the hot nebula. In 1980 Clayton reported that he had determined the isotopic composition of the ^{16}O-rich material. Searches are underway to try to identify some of the grains themselves.

The swirling nebula somehow never thoroughly mixed the ^{16}O-rich material while the planets and asteroids were forming. Remarkably, two or three reservoirs of the ^{16}O-poor gas were not even homogenized. Evidently the meteorites and moon rocks come from at least a dozen different, unmixed places in the solar system. It so happens that both the Earth and the moon are made of materials from the same part of the nebula, as are two types of meteorites. Clayton's data therefore suggest that the moon and the Earth have always been associated and that most of the material composing the moon certainly was not captured from far distant parts of the solar system.

If we now could just find where those moon- and Earth-like meteorites are coming from, we might learn the size of the Earth's zone of the nebula. These particular meteorites aren't exactly moonlike in their chemistry, so they don't come from the moon itself. The spectra of some main-belt asteroids look tantalizingly like those of the meteorites. Possibly those asteroids formed near Earth and were implanted into the belt long ago, or perhaps the meteorites come from small asteroids much closer to the Earth. Then Mars, which is between Earth and the main asteroid belt, could have formed from its own pot of nebular material and we would not have to hypothesize that the Earth's zone was so wide. These fascinating questions about the creation of the planets are being answered, thanks to measurements of some rocks that fall freely from the skies, of others that were returned from expensive lunar expeditions, and of still others to be returned one day from missions currently on NASA's drawing boards.

Meanwhile, lunar researchers have already thought of new experimental techniques to apply to the lunar rocks that will further constrain our hypotheses for the origin of that pale disk about which mankind has always wondered. After centuries, even millennia, of theorizing about how our moon was formed, we are within a few years—or a few decades at most—of finally knowing the answer.

26. This view of the receding planet Saturn, taken several days after Voyager 1's encounter, is one of the most beautiful pictures taken during the first two decades of planetary exploration. Note how the shadow of the planet's globe eclipses the rings. *Courtesy NASA*

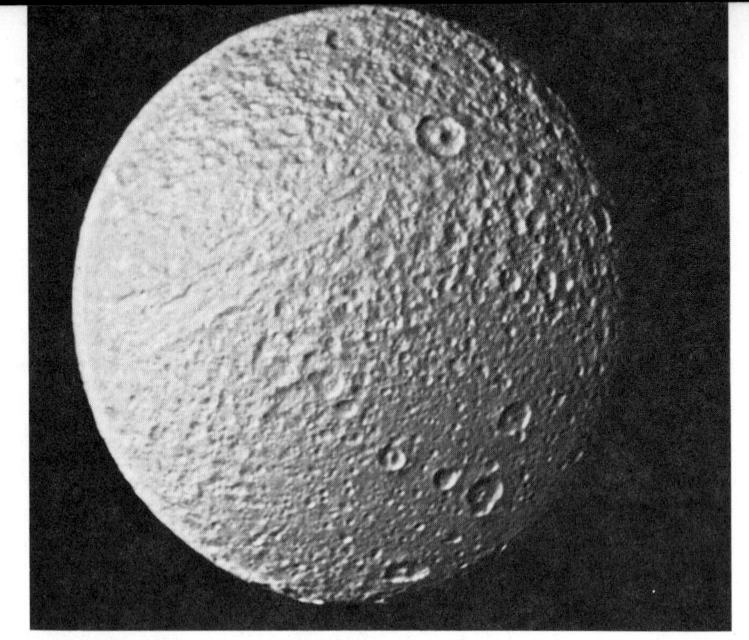

27. What does a body 1,000 kilometers in diameter, the size of the largest asteroid, look like? This Voyager 2 picture of Saturn's moon Tethys provides one answer. A small but prominent crater lies near the end of a system of deep canyons and chasms that girdles nearly three-quarters of the circumference of Tethys. *Courtesy Jet Propulsion Laboratory, NASA*

28. This Voyager 2 picture of Saturn's two-faced satellite Iapetus was taken on August 22, 1981. In a reproduction, it is difficult to capture the icy brilliance of the bright side of Iapetus (to the right) as well as the blackness of the dark side, which is blacker than tar. *Courtesy Jet Propulsion Laboratory, NASA*

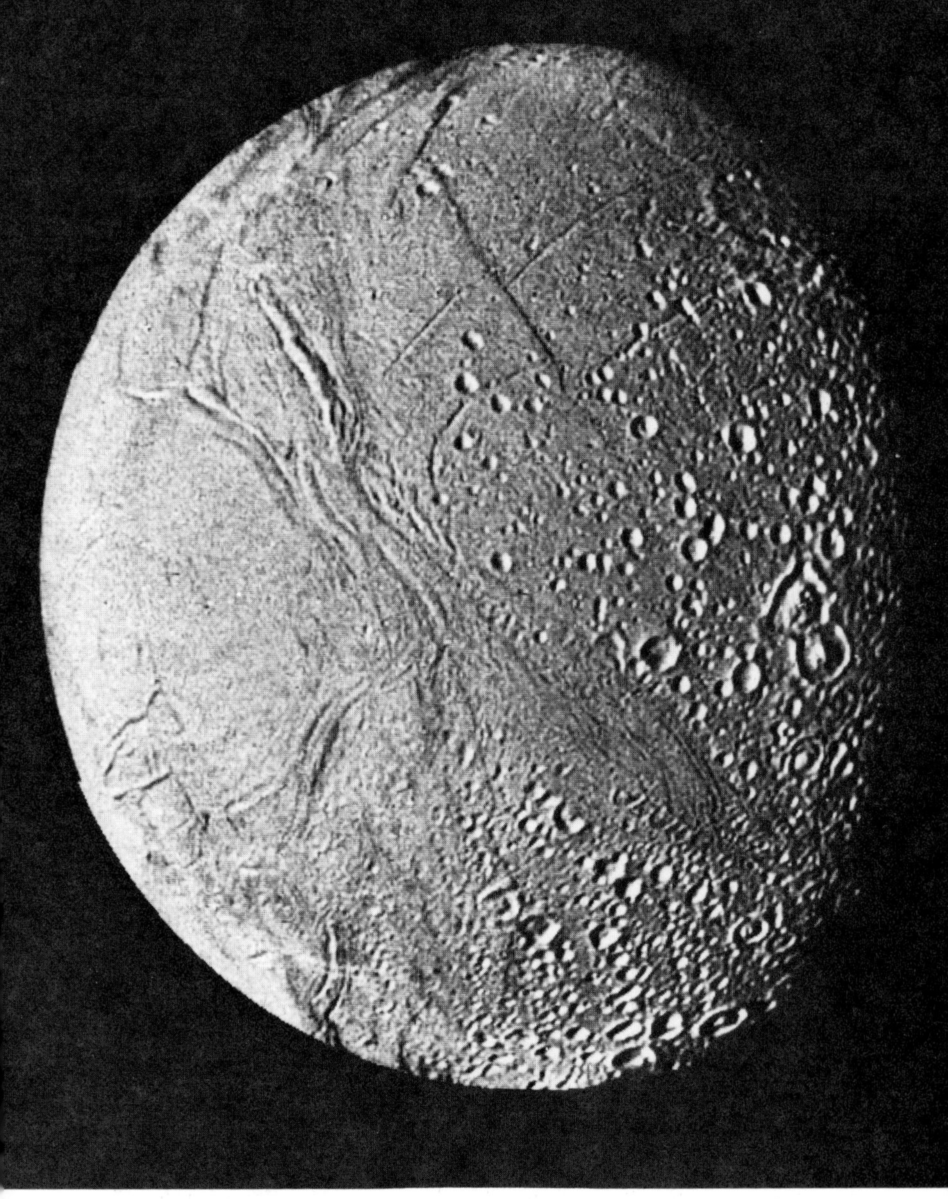

29. After Voyager 1's distant views of Enceladus showed it to be lacking in numerous deep craters, scientists began to suspect that it might be the geologically youngest of Saturn's icy moons. Nine months later, in August 1981, Voyager 2 took this picture of the remarkable satellite, revealing faults and some craters, but also vast craterless plains and grooved terrain. Scientists are still puzzling over what internal (or external) sources of energy can be keeping this small, cold world geologically active into modern epochs. *Courtesy Jet Propulsion Laboratory, NASA*

30. This view looking back at the translucent rings of Saturn was captured by Voyager 1 following its historic encounter with Saturn in November 1980. The shadow of the rings is visible on the overexposed globe of Saturn. The faint, narrow stripe just below the bottom, broad ring is the mysterious F-ring. *Courtesy NASA*

31. The splotches visible in the brighter parts of Saturn's B-ring are the famous spokes, first discovered by Voyager 1 in 1980. The A-ring is visible in the distance, to the upper right, in this Voyager 2 picture. *Courtesy NASA*

32. Uncountable ringlets and gaps in Saturn's middle B-ring are visible in this picture, which covers only about one-third of that ring. The picture was taken by Voyager 2 on August 25, 1981, shortly before the spacecraft's scan-platform malfunctioned and cut short investigation of the famous rings. *Courtesy NASA*

33. The rings seem strangely dim in contrast to the over-exposed crescent of Saturn, in this view. It is because we are looking down on the *shadowed* side of the rings, so only the more translucent parts of the rings show up as bright. This photograph was taken by Voyager 2 on August 29, 1981, shortly after engineers were able to free the spacecraft's scan-platform, which had become stuck shortly after ring-plane crossing several days earlier. *Courtesy Jet Propulsion Laboratory, NASA*

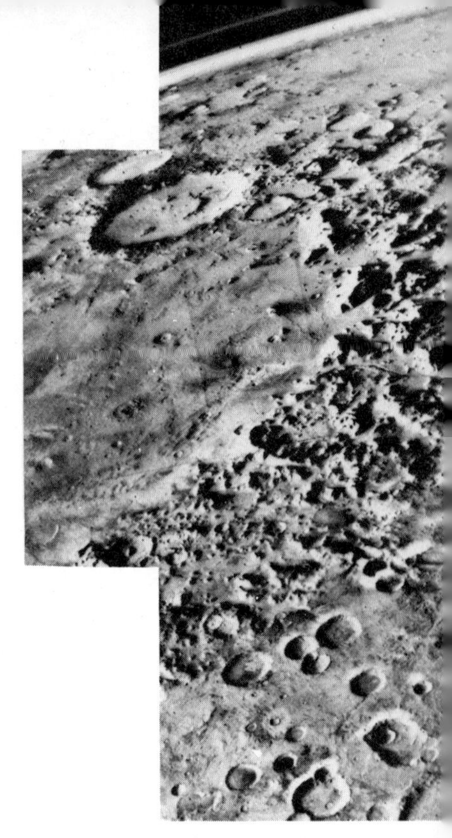

34. This oblique view across Mars toward the southern horizon some 19,000 kilometers away was obtained on the night of July 11, 1976, by one of Viking 1 Orbiter's two TV cameras. The picture is dominated by the greatly eroded rim of the huge Argyre Basin, the smooth floor of which is to the left. Above the horizon are layers of haze, 25 to 40 kilometers high. *Courtesy NASA*

35. This panorama of sand dunes, rocks, and boulders on Mars was taken by the Viking 1 lander about two hours after Martian sunrise. The boom that supports a small weather station cuts through the center of the picture. *Courtesy NASA*

36. Viking 1 Orbiter photographed this region of Mars, which apparently has been carved by large river channels. The ground slopes from west to east across the picture. The Lunae Planum highlands are to the west and the Chryse plains, in which Viking 1 landed, are to the east, well beyond the right-hand border. *Courtesy NASA*

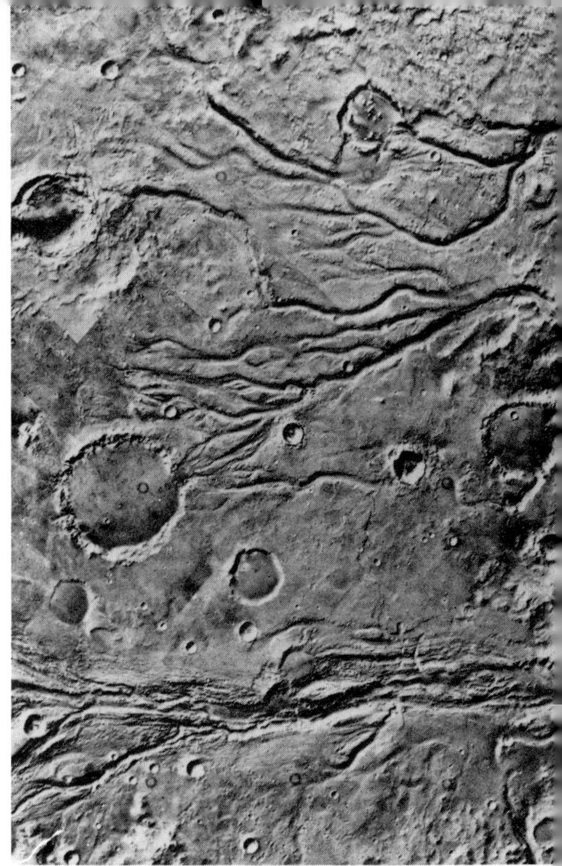

37, 38, 39. The Olympus Mons shield volcano on Mars is the largest volcano known on any planet. First named Nix Olympica ("Snows of Olympus") because of its frequent white appearance through telescopes from Earth, its true nature is hinted at in photographs taken as Mariner 6 approached Mars in 1969 (above). Nix Olympica is the whitish ring with darker interior just above the middle of the planet. Perhaps a hint of the summit crater can be seen, but the hints went unrecognized until a 1971 dust storm dropped beneath the flanks of the huge volcano, revealing its majestic topography and 500-kilometer dimensions to Mariner 9 (opposite, top). In the high-resolution view of the texture of the slopes of Olympus Mons (opposite, bottom; downslope is to the upper left), the trained eye of a photogeologist quickly recognizes the ridges, cracks, and the tongues of once-flowing material as characteristic of lava flows on the sides of Earth's volcanoes, such as Mauna Loa in Hawaii. *Courtesy NASA*

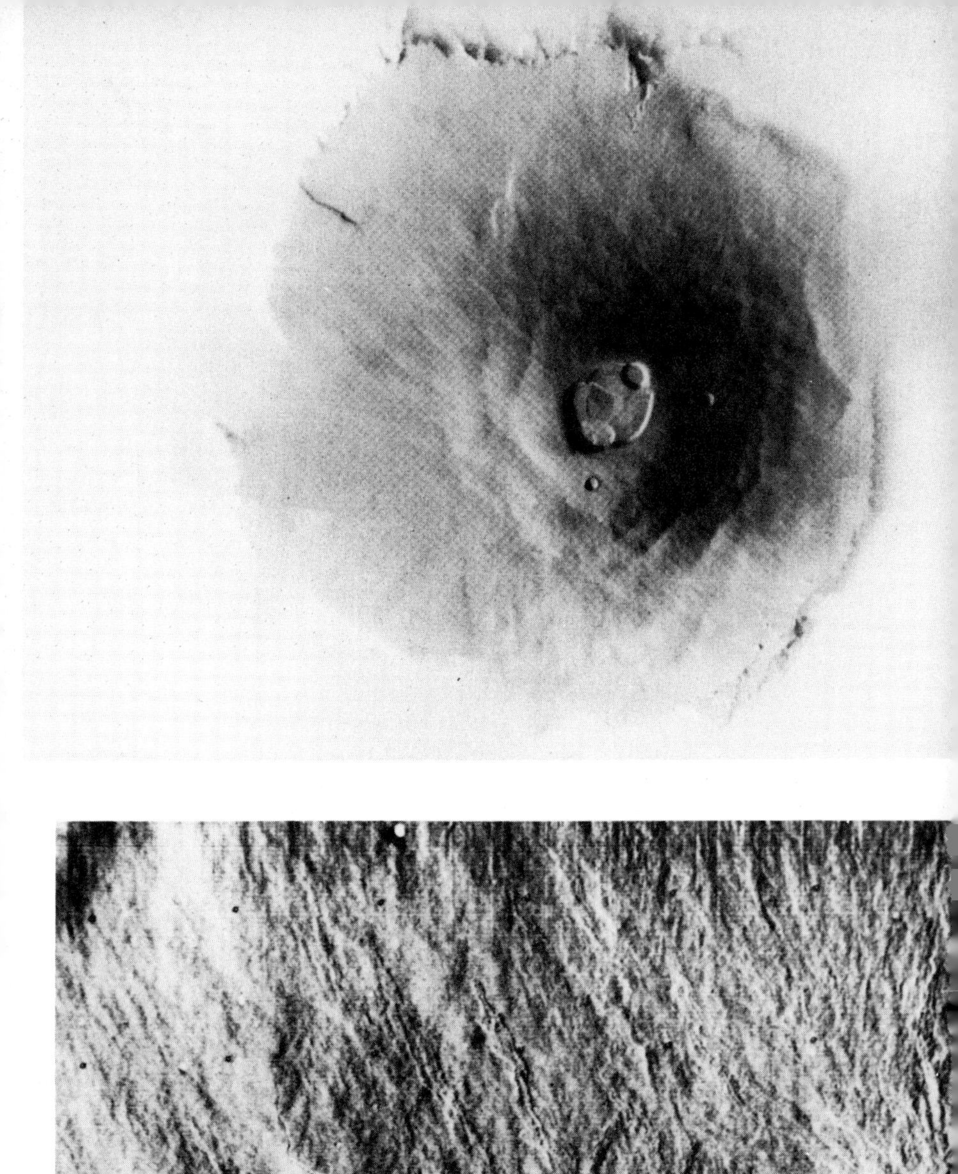

40. The surface of Mars is covered with rocks near the Viking 2 lander. Many of the stones have small holes, or "vesicles," in them. The angular rock in the right foreground is about 25 centimeters across. *Courtesy NASA*

41. The flower-petal-like deposit around this crater is characteristic of many Martian craters. Some scientists call them "splosh craters." Probably water-ice-rich Martian ground responds differently to impacts than does the dry lunar or Mercurian ground. Perhaps the atmosphere of Mars also helps to control the shape of crater ejecta blankets on Mars. *Courtesy NASA*

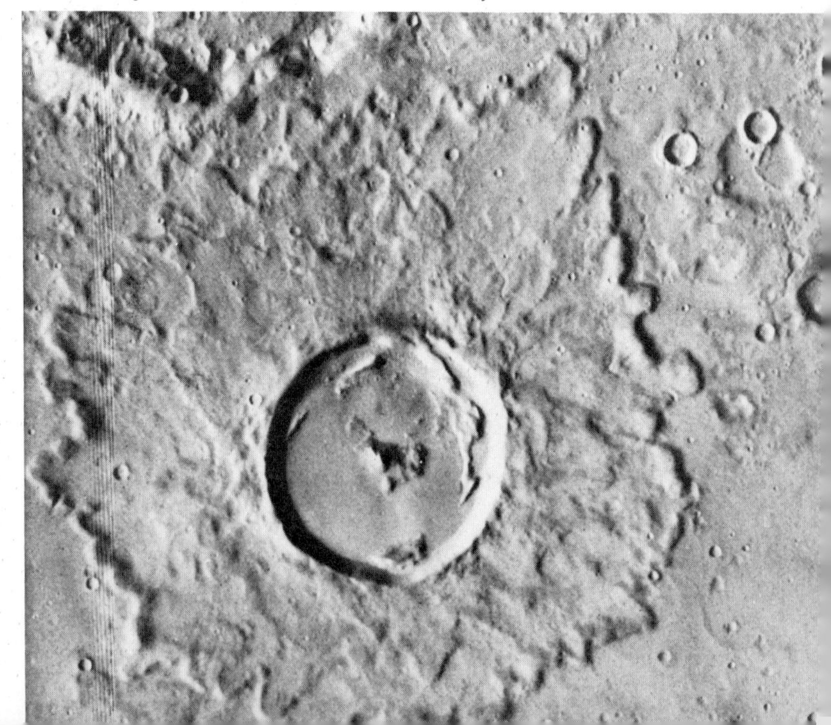

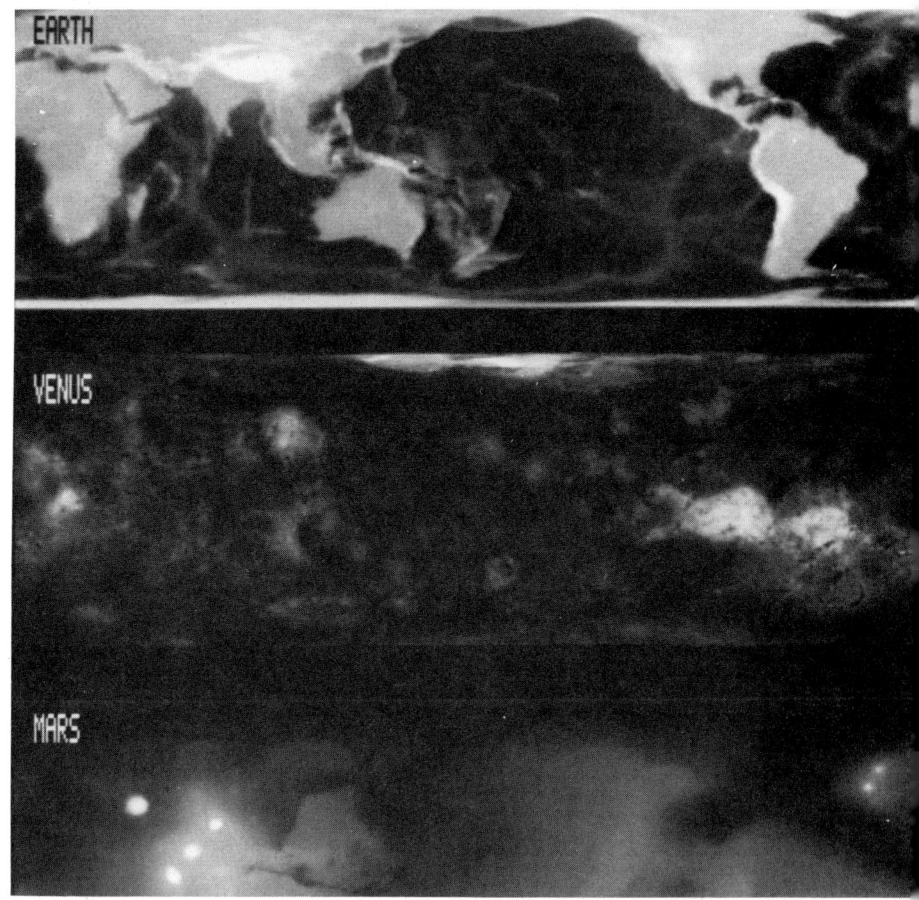

42. There are great differences in topography on the Earth, Venus, and Mars as shown in these three maps, all plotted on the same map projection. The darkest regions are the lowest; brightest regions, the highest. Sea-floor topography is shown for the Earth; note the mid-Atlantic ridge. Venus topography was mapped by Pioneer Venus. Note six high volcanoes on Mars and the dark slash of the Valles Marineris canyon. The maps were all computer processed at Washington University by Raymond Arvidson, who kindly made them available for this book.

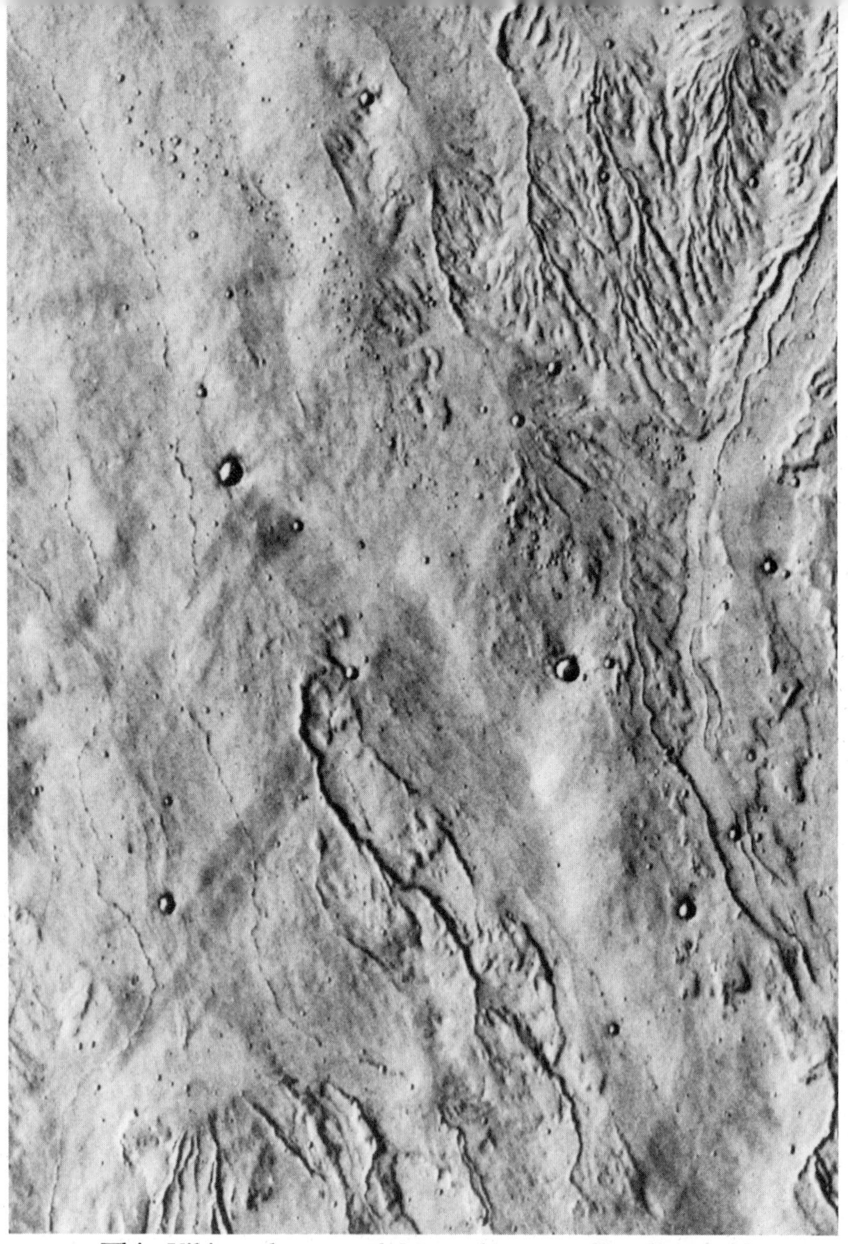

43. This Viking close-up of Mars shows small valleys, ridges, and curving channels that look superficially like stream valleys and hills shaped by rainfall run-off in aerial photographs of Earth. Detailed analysis of the Martian topography has revealed, however, that other processes (sapping by underground aquifers, volcanic processes, and subsequent "exhumation" of once-buried terrains) are probably responsible for the ubiquitous networks of small channels on the Red Planet. *Courtesy NASA*

44. This aerial close-up of Martian canyonlands shows jumbled chaotic terrain at the foot of enormous eroded cliffs. *Courtesy NASA*

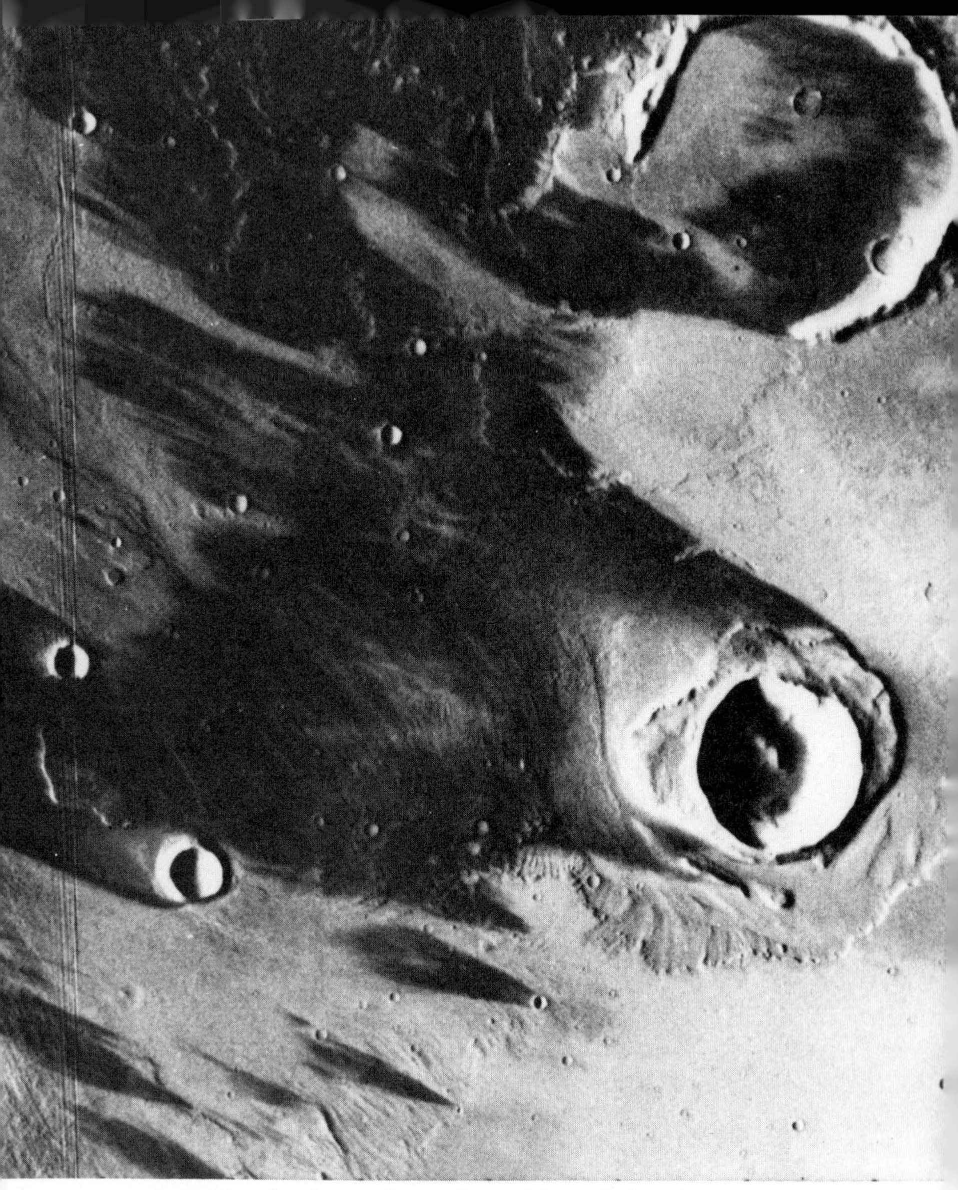

45. Dark splotches and streaks on Mars change from year to year and season to season due to the shifting winds. Thin deposits of light-colored dust, left by annual global duststorms, are subsequently stripped away, revealing darker ground beneath. Craters are obstacles in the wind flow, creating downstream turbulence, just as some buildings do in Earth's cities. That is why streaks are often associated with craters. From the changing orientations of the streaks, scientists were able to monitor Martian winds during the four years of the Viking mission. *Courtesy NASA*

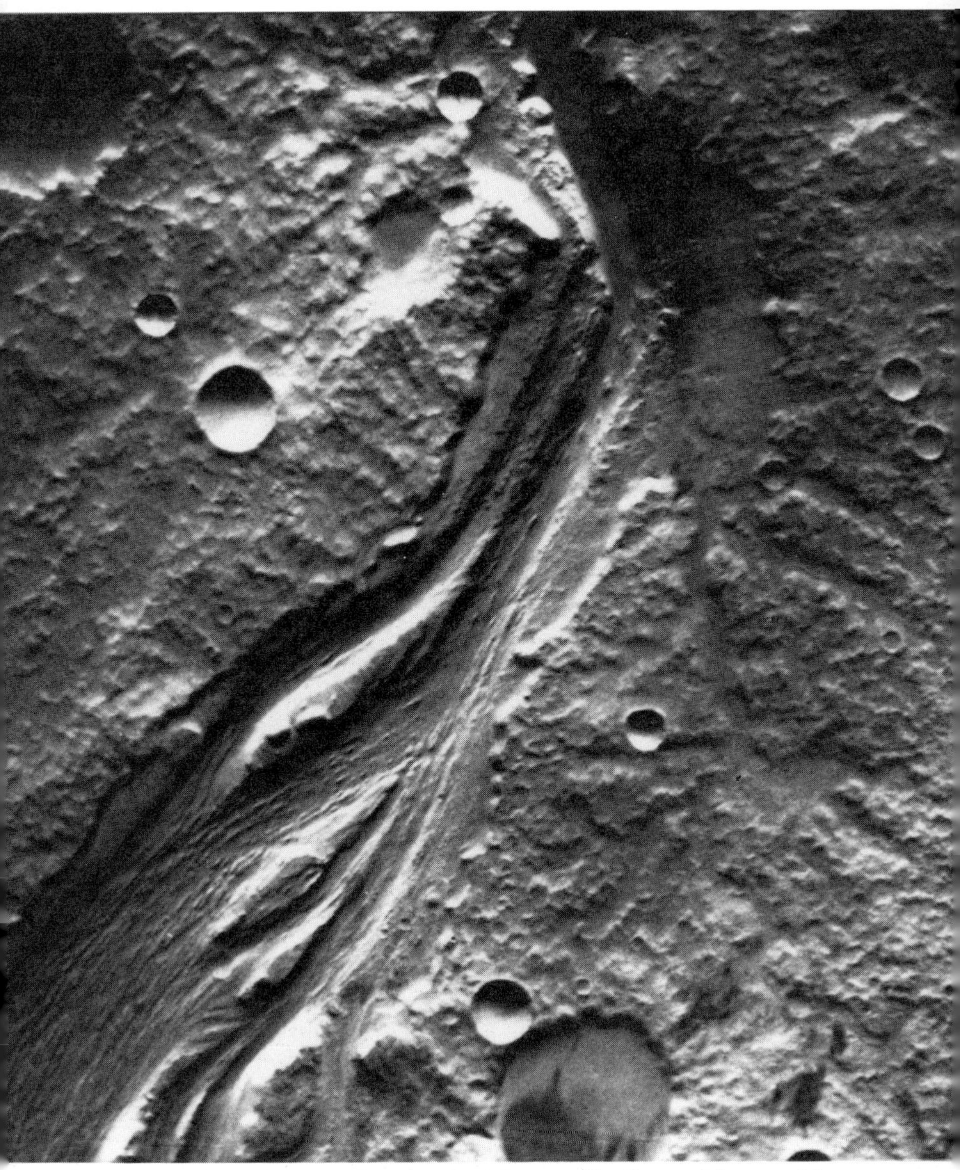

46. Viking close-up of a large "river valley" on Mars, with several "islands." The tiny, elongated hills on the floor of the channel may be a clue to its origin. Perhaps they are stream-bed features, subsequently eroded by windblown dust into elongated yardangs like those in the driest deserts on Earth. Or perhaps they are more akin to drumlins, which are formed on Earth by glacial processes. *Courtesy NASA*

47. The inset shows the larger Martian moon, Phobos, with its giant crater, Stickney. The close-up reveals an intricate pattern of grooves and crater chains, some of which criss-cross each other near the bottom of the picture. Although the grooves are clearly related to the Stickney impact, which almost fractured Phobos into a swarm of smaller moonlets, the subsequent dynamical evolution of Phobos has also left its mark on the pattern we observe today. *Courtesy NASA*

9

Galileo's Worlds of Ice, Rock, and Sulfur

A year and a half ago there was no reason to expect the new and the unusual in the visible spectra of the Galilean satellites of Jupiter. They were, after all, discovered in January, 1610, and they have been easy objects of spectroscopic studies for more than 50 years. . . . In early summer of 1972, we discovered an anomalous brightness in the spectrum of Io near the sodium D-lines. . . . In all probability this discovery will prove a useful tool in probing both the Jovian environment and the conditions on this fascinating satellite.

—*Robert A. Brown, 1974*

The first time many Americans ever heard of a place called Io was when they tuned in the evening news broadcasts on Monday, March 5, 1979. Bradford A. Smith, leader of the team of scientists studying new pictures from the Voyager spacecraft, showed a bizarre view of an orangish disk, splotched red and white, with what looked like yellow pustules and black scabs. This was Io,* one of Jupiter's largest moons, which Smith

* Perhaps many Americans subsequently forgot about Io due to its confusing name. My colleagues have strong opinions about how to pronounce it. I say "ee-oh" and "eye-oh" interchangably. In Greek legend, Io was a mistress to Zeus (Jupiter).

likened to an unappetizing pizza. Within a week, baffled scientists realized what Voyager's pictures and other data implied: Io was a world of fire and brimstone, sulfur dioxide snows, immense fountaining volcanoes, all trapped in a whirling magnetic maelstrom and bathed in deadly radiation.

For a few scientists, however, Voyager's historic discoveries about Io only crystallized what they had long felt in their bones —that Jupiter's innermost major moon might be the most remarkable planet in the solar system. For years such satellitologists as Dale Cruikshank, Torrence Johnson, and David Morrison loved to tell about just how *weird* Io was. From the time of Galileo's discovery of Jupiter's four large moons (since called the Galilean satellites), Io's special place in the solar system has been apparent: it is the closest world to the largest planet. Jupiter looms in its sky larger than the bowl of the Big Dipper; at night the polychrome, belted giant is brighter than 100 of our full moons! Io's proximity to Jupiter has profoundly dominated its fate, as planetary scientists started to learn during the two decades preceding Voyager.

The first indication that Io was special came quite early in the history of radio astronomy. Everyday objects and planets alike, due to their temperature, emit radiation throughout the spectrum, including wavelengths invisible to the eye. Hot objects, like heating elements on a stove, glow visibly. At cooler room temperatures and those that prevail on most planets, such thermal emission is strongest at longer infrared and radio wavelengths. In the 1950s, bursts of radio activity were detected from Jupiter, far more intense than its normal thermal radiation, making the planet the most powerful source in the sky of radio waves 10 to 30 meters long. In the early 1960s it was found that this nonthermal radio noise was controlled by the orbital position of the satellite Io. The bursts were heard when Io was in certain parts of its orbit, but generally not at other times. How could little Io be responsible for the noisiest static in the solar system?

About the same time, two Arizona graduate students in planetary studies made observations of Io that have been controversial ever since. Alan Binder and Dale Cruikshank measured the brightness of Io after it emerged from eclipse in Jupiter's shadow. (Io orbits Jupiter every two days, spending about 2 hours in its shadow before reemerging into sunlight.) They found that Io was brighter for about 15 minutes after eclipse than it had been before entering the shadow. Perhaps a thin frost had been deposited on Io's surface during the brief "night," the students reasoned. They concluded that Io probably has an atmosphere of some sort. However, in 1971 it was proved that Io has a negligible atmosphere, with a surface pressure less than a millionth that on Earth. (We now think it may be as little as a millionth of a millionth of Earth's atmosphere, virtually a vacuum!)

Other astronomers tried to observe the "Binder-Cruikshank effect" themselves and were mainly unsuccessful. A few occasionally did see anomalous brightenings, but generally not at all. Satellite experts split into two groups: those who conservatively regarded the detections as erroneous and those who, with cautious optimism, believed the brightenings were sporadic but real. At the same time, there were suggestions that Io's "strong orange" color, as Gerard Kuiper had described it in the early 1950s, might be variable over the years.

Io's surprises were just beginning. In the early 1970s, Robert Brown—then at Harvard University—detected a glowing cloud surrounding the planet. The glow is light emitted by atoms of sodium—a soft, combustible metal that, in combination with chlorine, forms innocuous table salt. How could Io have a metallic atmosphere, scientists wondered. Actually sodium is an unusually strong emitter and easy to detect even when present in small quantities. But it was a clue that other atoms might be there, too. Astronomers began searching for them using more sensitive telescopic techniques. It became clear that Io was imbedded not just in a cloud, but in a torus or gigantic donut-shaped ring of atoms and ions composed of potassium, oxygen, and sulfur as well as sodium. The *atoms* apparently surround Io's orbit all the way around Jupiter. The ring of *ions* is also the same size as Io's orbit, but tilted about 10 degrees to it, the same tilt as the equator of Jupiter's magnetic field. An ion is an atom that has been stripped of one or more of its outer, negatively charged electrons, which normally counterbalance the positive charge of the nucleus in an atom. So an ion is a positively charged atom, easily dragged, accelerated, and captured by a magnetic field like Jupiter's. Evidently Io is imbedded in a tubular "plasma"—a cloud of hot ions and electrons—trapped and corotating every 10 hours with Jupiter's magnetic field.

Where could the neutral atoms and corresponding ions be coming from? Their association with Io's orbit pointed the finger straight at Io. It would be odd for Io's surface to be covered by sodium- and sulfur-rich material, yet apparently it was. The spectrum of light reflected from Io was found to be similar to the spectrum of sulfur and sulfur compounds. And Io's spectrum showed no trace of water-ice, conspicuous in spectra of other Galilean satellites. Tentative hypotheses were advanced to explain the proposed enrichments of the unusual atomic species on Io's surface (dry, salt-encrusted lake beds was one idea). Theorists worked on the problem of how to get the material off the surface of Io, ionized, and into orbit around Jupiter in the torus. Thinking was still quite confused, although partly on what turned out to be the right track, by the time Voyager 1 encountered Jupiter in 1979. Satellite specialists were eagerly anticipating pictures of Io, for it was already clear that Io was a special place.

As Voyager approached Jupiter, yet another remarkable observation of Io was made from Earth. One night Io was found to brighten for several hours by a factor of 3 to 5 at thermal infrared wavelengths, then fade again. Assuming they had observed a "hot spot" on Io's surface, the observers calculated that it must have been at the melting temperature of lead and about 100 kilometers across! Could it be an eruption of molten magma? "Although it is not possible to definitively rule out this explanation," the scientists wrote, "we consider it to be unlikely by analogy with our experience with objects in the inner solar system. No such internally generated hot spot . . . has ever been observed on global infrared surveys of Earth, the moon, Mercury, or Mars." Volcanism on Io more active than on Earth was simply too farfetched to believe, so the scientists concocted several nonthermal explanations for the infrared brightening.

It is fascinating that most pre-Voyager discoveries concerning

Io might have been made years, or even decades, earlier. In fact, abundant evidence for thermal anomalies existed in data records extending back for a decade. The sodium emission could have been measured as early as 1910, and actually *was* measured, but overlooked, many years before Robert Brown's discovery. Nobody imagined that Io could be as weird a body as it kept revealing itself to be. One wonders how much the flights of the Pioneers and Voyagers inspired ground-based observers to work harder and try to "beat" the probes to their target. The most remarkable discovery of all about Io was made during just such a race, one that came right down to the wire. In this case, it was not sophisticated instruments on big telescopes that bested Voyager. The tools were simply the unadorned brainpower of three theoretical physicists who tackled the problem of the origin and evolution of Io as Voyager 1 closed in on its target. Theoretical ideas that had their genesis in the turn-of-the-century work of Sir George Darwin were further developed and won the race in a stunning photo finish.

Stanton Peale, a physics professor in Santa Barbara, California, was one of the scientists who studied Io. He is an athletically built man with a precise way of speaking—and thinking. He teamed up with two other West Coast theoreticians, Ray Reynolds and Patrick Cassen, to study tidal processes affecting the Earth and the moon. In the summer of 1978, they wondered if their ideas about lunar tides might be applicable to Io or to the other Galilean satellites.

Tides result when two large bodies are near each other. When the moon is overhead, it pulls harder on the side of our planet facing it than on the other side, which is farther away. This differential pull elongates the Earth in the direction of the moon so that the ocean flows and bulges up even more towards the moon. Io is so close to mighty Jupiter that Io bulges between 5 and 10 kilometers! Once Io may have been spinning rapidly about its axis, with huge tides raising and lowering its surface every day or so. Were Io perfectly elastic, it could have spun on, eternally flexing itself with no problem. But like other planets, it is made of rocks and liquids that respond sluggishly to such manipulation, dissipating heat in the process. As a result, Io's lagging tidal bulge was swept on beyond the line joining Io and Jupiter, just as peak ocean tides occur after the moon has been overhead.

Jupiter's gravity tugged back at the tilted, oblong Io, tending to pull it back into line and slow down Io's spin. Just such effects, operating long ago, despun all the major satellites in the solar system, including the moon, so that ever since they have kept their same faces—more or less—toward their parent planets. The "more or less" depends on the elongation or eccentricity of a satellite's orbit; it is a crucial qualification for Io. A satellite in an eccentric orbit moves around faster when it is close to its planet than when farther away. However, it spins about its axis at a constant rate, so its orbital revolution sometimes carries it ahead of its spin while at other times it lags behind. Therefore, the tidal bulges oscillate back and forth across the satellite's surface, in addition to rising and falling in response to the satellite's changing distance from its planet. Friction resulting from the tidal flexure generates heat within the satellite.

Ordinarily, Io would have responded to Jupiter's forces by moving away and evolving into a more circular, distant orbit, soon diminishing the tides and heating. But Io is trapped in its present orbit, and its orbital elongation is continually reinforced by neighboring Europa. This is due to a curious relationship between the orbits of Io, Europa, and Ganymede, first explained around the time of the French Revolution by the Marquis de Laplace. Europa takes just about twice as long as Io to revolve about Jupiter. Similarly, Ganymede's period is twice Europa's. This "resonant" relationship builds up strong gravitational forces that have held Io captive, perhaps for billions of years. Thus tidal heating of Io's interior has never ceased.

Peale, Cassen, and Reynolds calculated that such heating within Io would be at least three times larger than that produced by decay of long-lived radioisotopes—the source of heat generated within the interiors of the Earth and moon. Analysis of moonquakes suggests that the moon is partly molten near its center. So the extra heating of Io, which is the same size as the moon, must surely have melted its center. But Stan Peale and his colleagues realized that once Io had a molten core, there would be even more heating of its solid mantle. More melting would occur, leaving an even thinner solid shell, which would be still more efficiently heated. And so on. There would be runaway melting of Io until most of its interior was molten, leaving a thin solid crust only 20 kilometers thick. *Thirty times* as much heat would be generated in a thin-shelled Io as within the moon!

The researchers wrote a manuscript entitled "Melting of Io by Tidal Dissipation" in January 1979. The magazine *Science*, which often takes 5 or 6 months to publish an article, rushed the report into print in 6 weeks, just 3 days before Voyager arrived at Jupiter. The dramatic conclusions were these:

> [Our] calculations suggest that Io might currently be the most intensely heated terrestrial-type body in the solar system. . . . One might speculate that widespread and recurrent surface volcanism would occur, leading to extensive differentiation and outgassing. . . . Voyager images of Io may reveal evidence for a planetary structure and history dramatically different from any previously observed.

Needless to say, Voyager provided stunning confirmation. Human intellect ran a race with a multi–hundred-million-dollar spacecraft mission and won! If anything, Peale *et al.* underestimated what could be happening on Io. Of course, the nine active volcanoes observed to be spewing powder, gas, and snows 50 to 300 kilometers above Io's surface, and as much as 500 kilometers off to the side, dwarf Earth's active volcanoes, including mighty Mount St. Helens.

Perhaps the most fundamental quantity that measures Io's dynamic activity is its "heat flow." As I discussed in Chapter 6, a planet's internal heat governs its geological activity. The average amount of heat flowing to the planet's surface and radiating to space must be the same as the internal heat production; otherwise the planet would heat up or cool off until it again came to equilibrium. Apparently most of Io's heat is escaping from "hot spots," many associated with visible volcanic craters or fissures. A heat-sensitive instrument aboard Voyager mapped some hot regions and measured their temperatures, ranging from room temperature—which is hot for Io's frigid environment—to 600 K (620°F). The total heat flow for several days near the Voyager 1 encounter was several times greater than even the maximum predicted by Peale and friends! Was something else heating Io in addition to tides? More likely Io was just a bit more active than usual. After all, one of the hottest spots in March 1979 was the volcano named Pele, whose gigantic fountain was inactive four months later when Voyager 2 flew by.

Some astronomers reexamined Earth-based measurements of Io extending back to 1969, while others began monitoring Io anew, using telescopes atop the 13,800 foot summit of Mauna Kea on the "Big Island" of Hawaii. These observations, over a dozen years, agree with Voyager's data: on average, thirty times more heat is flowing out through Io's surface than through an equivalent area on Earth. Only a few of the most geothermally active places on Earth, with hot springs, geysers, and volcanoes, match the *average* activity everywhere on Io! Perhaps the 1970s were an unusually active decade for Io.

Voyager's pictures and data finally have revealed the remarkable nature of Io and explained the earlier mysteries. Io's surface is continually renewed by processes reflecting its restless internal activity. Fissures crack open over new hot spots and towering plumes erupt, blanketing the landscape with sulfurous dust. Immense floods of molten black sulfur pour out across hundreds of kilometers of Io's surface, changing to brown, red, then orange, and finally yellow as they cool. Io's equatorial and temperate zones contain hundreds of great volcanic calderas that are more than 10 miles across: these collapsed craters dwarf their terrestrial counterparts and represent the more quiescent, mature stage of Ionian volcanism. Great eroded mountains tower miles into the sky, possibly remnants of ancient Earth- or Mars-like silicate volcanoes.

Io's subterranean cauldron notwithstanding, the height of the towering plumes strains credibility. Velocities approaching 1 kilometer per second far exceed what can be generated in the throat of a terrestrial volcano. Although we cannot peer beneath Io's surface, radically different physical and chemical processes must be occurring there. Probably Io has a crust of solid sulfur, sulfur oxides, and perhaps sulfur-enriched silicates. Sandwiched between it and a deep rocky mantle of familiar silicate composition are lakes or oceans of liquid sulfur compounds. Especially hot silicate magmas emerging from Io's interior vaporize the sulfurous materials. Apparently the plumes are powered by the explosive expansion of vaporized sulfur and/or sulfur dioxide, like steam venting from a pressure cooker.

The puzzling Binder-Cruikshank posteclipse brightening is probably, after all, a real but irregular phenomenon. Cosmochemist Fraser Fanale of the University of Hawaii believes that thin, ephemeral sulfur dioxide frosts sometimes can be deposited on the warmer parts of Io's surface during eclipses, but only when more sulfur dioxide gas than usual is vented by nearby plumes. Fanale calculates that such sporadic frost deposits would sublime away in about 15 minutes following reemergence of Io into sunlight, which is in agreement with the observations.

Apparently, volcanic fountains help replenish Io's giant torus with atoms. While even the most powerful fountains rarely reach Io's escape velocity, at least some plume material gets into the magnetosphere and is swept away from Io. Even a tiny fraction of such volcanic emanations would profoundly affect Io's torus. Most atoms in the torus have been blasted off Io by ions already in space. What powers this "chain reaction"? Imbedded deep within Jupiter is the dynamo that generates its magnetic field, which therefore whirls around at Jupiter's 10-hour spin rate, sweeping past Io at 50 kilometers per second. Thus ions caught up in the field slam back into Io at tremendous velocities. Surficial atoms are bounced, or sputtered, directly into space, when Io's transient atmosphere of sulfur dioxide is thin enough. At other times, the energetic ions knock atmospheric sulfur dioxide into space. The ultimate energy for this atomic sandblasting of Io is the immense rotational energy of Jupiter itself.

Theories developed a decade ago to explain Io's peculiar control over Jupiter's radio static predicted that Io was connected to Jupiter by a "flux tube" of currents, generated by Jupiter's magnetosphere sweeping past the satellite. Voyager 1 missed its targeted flight through the putative flux tube, but it flew close enough to confirm that the tube exists and was continuously carrying about 5 million amperes of current. That is a hundred times the peak current carried for a tiny fraction of a second in a lightning stroke. Dazzling phenomena had been expected in Jupiter's atmosphere at the "foot" of the flux tube, but they were overwhelmed by a brilliant display of "northern lights." The Jovian aurorae are due to precipitation of particles along magnetic lines of force joining Jupiter's polar regions not just with Io (the flux tube) but with Io's entire plasma torus.

Perhaps more than any other body in the solar system, Io has brought together the diverse disciplines in planetary science. No specialist can understand this multifaceted world in isolation. Auroras on Jupiter are due to volcanoes on Io. The volcanism is caused by esoteric numerical orbital relationships of three satellites (Io, Europa, and Ganymede), mediated through the unfamiliar chemistry of sulfur. So intense and unusual are Io's interactions with Jupiter that they are manifested throughout the elecromagnetic spectrum, from the extreme ultraviolet glow of the plasma torus through Io's shocking visible colors and hot-spot thermal radiation to kilometer-wavelength radio emission. Io is indeed the strangest place in the solar system discovered so far.

It has often been written that Jupiter and its system of satellites is a miniature solar system. Revolving in circular, coplanar orbits around a massive central body, Jupiter's moons superficially resemble planets orbiting a star. But the analogy is much more fundamental. For example, Jupiter is actually radiating more energy than it gets from the sun, like a very faint star. The bulk densities of the Galilean satellites decrease with increasing distance from Jupiter—Io and Europa have dense, rocky interiors, whereas Ganymede and Callisto are icy—just as the planets vary from metal-rich Mercury through rocky Earth to the lighter outer planets. Moreover, the orbital resonances among the first three Galilean satellites, which have had such a profound effect on Io, have counterparts among the asteroids and the planets. If the Jupiter system is truly a "second" solar system, we can test our developing theories for planetary origin on Jupiter's moons.

Jupiter probably formed in the center of a large, rotating disk of gas and dust. This massive Jovian "nebula" was far enough from the sun that it was probably of the same composition as the sun itself, unlike the inner planets, which were depleted of the lighter, more volatile gases due to the heat in the inner solar system. Calculations show that as Jupiter contracted, it became a luminous star for a while, perhaps a thousandth as bright as the sun. This was sufficient to drive volatile compounds away from the smaller bodies (planetesimals) in close-in orbits that were accumulating or accreting into Io and Europa.

Ganymede and Callisto formed from planetesimals farther away from Jupiter, so they received their full cosmic complement of low-density water-ice, accounting for one-third to two-thirds of their masses. Accretionary impacts of planetesimals may have warmed the forming moons and helped convert much of the ice within their exteriors into enormous oceans of liquid water. The surfaces of Ganymede and Callisto, radiating heat into space, froze at once. As the bodies cooled, their ice crusts thickened at the expense of their immense mantles of liquid water. Rocky materials, three times denser than water, sank toward the center of Ganymede (probably of Callisto too), just as pebbles fall to the bottom of a lake. After the satellites had formed and Jupiter cooled, the dominant heating within the moons came from radioactive decay of uranium, potassium, and thorium within their cores. (Radioactive elements occur in rocky minerals, not in ices.) Ganymede is so large (between the sizes of Mercury and Mars) that its rocky core alone is the same size as the moon! Because of its larger, radioactively warmed core and its tidal interactions with Io and Europa, Ganymede's ice crust probably remained thin for a longer time than did Callisto's. Eventually the crusts thickened on both bodies and their mantles probably became frozen solid. But it is not known how well icy crusts would have conducted heat from a liquid mantle to the surface. If the mantles contain salts and other impurities, like a kind of planetary antifreeze, their freezing temperatures would be depressed. So Ganymede and conceivably Callisto could still have liquid mantles today.

Callisto is smaller, less rocky, and farther from Jupiter than Ganymede. It formed more slowly and it has negligible tidal heating. For all these reasons, Callisto probably never was as warm as Ganymede. Rock and ice may not even have substantially segregated within it. Certainly it cooled and froze more rapidly and completely. The geological processes on Ganymede and Callisto are determined by the properties of ice and water, not rock and magma as for planets explored by earlier spacecraft. Telescopic spectra show that Ganymede, especially, is covered with abundant water-frost or water-ice. So as Voyager's cameras homed in on the outer moons, scientists were looking to see what the geology of ice would be like and whether there might even be ice-water analogs of continental drift to study.

The pictures were not disappointing. While frozen Callisto superficially resembles the moon, with craters upon craters, closer study reveals the telltale traits of an icy crust. For one thing, there are no tall mountains on Callisto. Its rare large craters lack towering ramparts and have only the barest hint of shallow topography. They have been almost completely erased and remain dimly visible on the freshly recratered terrain as brighter circular patches termed "palimpsests." The largest "crater" of all on Callisto is named Valhalla. It is a large palimpsest nearly the size of Texas, encircled by regularly spaced concentric rings, composed of cliffs, ridges, and broad, twisting canyons. The entire bull's-eye measures between 3,000 and 4,000 kilometers across, about two-thirds the diameter of Callisto itself! Yet the scarp rings—like ripples from a stone dropped in a pond, suddenly frozen—are all shallow and Valhalla's original crater walls have vanished. Apparently Callisto's topography has "melted" away. Over long durations, this is the way an ice crust should behave, much like "Silly Putty": both the Putty and ice are brittle when hit by a hammer, but they flow like viscous molasses given enough time.

It is not certain why Callisto (and Ganymede, too) is so deficient in giant craters. Perhaps many of the larger impacts occurred when Callisto's crust was very thin so that the craters vanished soon after forming. Or perhaps many large craters faded away over the ages, without leaving visible palimpsests. Some scientists think that the population of projectiles that have struck Callisto over the eons lacks large bodies, with the planetoid that formed Valhalla being a freak exception. Icy comets, newly derived from their home on the outer fringes of the solar system, are thought to be responsible for 60 percent of the craters on Callisto. Perhaps giant comets are very rare, and the large projectiles that have cratered the moon and terrestrial planets are uniquely characteristic of asteroids.

As scientists had predicted, Ganymede's landscape turned out to be much more diverse than Callisto's. An ancient epoch of geological upheaval is evident, possibly akin to plate tectonics, now "frozen-in" (see Chapter 12). The topography on much of Ganymede is bewilderingly complex, unlike anything seen on other planets. Large regions are crisscrossed by astonishing, haphazard patterns of lengthy valleys and ridges. From afar, they look like grooves, as though a mischievous Ganymedian giant had dragged a rake across the icy landscapes. The official term, grooved terrain, is surely a trivializing misnomer for such rugged icy domains, which are a couple of thousand kilometers long, each containing dozens of parallel valleys, thousands of feet deep and hundreds of kilometers in length, interspersed by Appalachian-like ridges.

The grooved terrains form great angular stripes and bands, separating the larger, polygonal, continent-sized cratered regions. The largest of the latter, called Galileo Regio, is a roughly circular, blackish "cap" on Ganymede's northern hemisphere. It is similar in size and shape to Antarctica. Some cratered polygons are crossed by parallel, curving ridges and gorges, which resemble parts of the bull's-eye pattern surrounding Valhalla on Callisto. Many Ganymedian palimpsests are evident as well. In contrast, the grooved bands, and some smooth ones too, are icy white and only sparsely cratered. Evidently Ganymede was once a Callisto-like world, but its dirty ice-rock crust fractured into pieces. Fresh water oozed up between the cracked plates and froze. Somehow the ice subsequently became "grooved."

Something had to give for the original crust of Ganymede to split apart and make way for the extensive grooved terrains. There is only so much surface area on a globe. Perhaps Ganymede expanded somewhat as its internal temperature changed. Some phase changes between water and the various forms of crystalline ice cause expansion, as anyone knows whose water pipes have frozen in the winter. But Ganymede could hardly have expanded its diameter by 40 percent, which would be necessary to accommodate the extensive bright areas. The Earth's crustal plates crack and continents drift, but room is made for new ocean floor at the expense of the older plates, which are submerged and forced into the Earth's mantle in a process called subduction. No evidence for subduction exists on Ganymede. Perhaps large peripheral parts of the great cratered continents fractured, foundered, and sank into the interior of Ganymede and were replaced by the bright, fresh ice. That would account for the fact that few of the separate polygons can be fit snugly back together like a jigsaw puzzle.

On the other hand, some scientists who have studied Ganymede believe that its crustal expansion has been minimal: just enough to form cracks, faults, and sunken valleys. Water then gushed out through the cracks and flooded the peripheries of the cratered lands which, these scientists believe, remain largely intact below the thin veneer of fresh ice. Did Ganymede have a violent period of crustal upheaval in an aquaeous version of plate tectonics? Or did it have a more staid history of watery volcanism along great fissure cracks in its originally Callisto-like surface? Either way, the sparsity of craters on much of Ganymede tells us that its surface was changing long after Callisto had stabilized. (Jupiter's huge gravity field should have concentrated more stray interplanetary projectiles onto Ganymede than onto Callisto, yet Callisto's surface has collected many more craters, and, therefore, must be more ancient.)

Finally, Ganymede's interior cooled, its crust thickened, and its surface became immobilized—a tableau of arctic plate tectonics. One final event shattered Ganymede's tranquility: a rare giant impact on Ganymede's southern hemisphere created a basin, Gilgamesh, which is remarkably like those on the moon and Mars. No palimpsest, Gilgamesh testifies that a thick crust of rigid, cold ice can mimic a rocky crust. We are reinforced in our realization that most cratering of Ganymede and Callisto occurred in an earlier epoch when the interiors of these bodies were much warmer and their crusts more plastic than they are currently.

The origin of the grooves in Ganymede's fresh, icy zones remains a puzzle. One idea is that they reflect the response of thin sheets of ice to tensional stresses due to powerful, ancient forces within Ganymede. According to the hypothesis, once a crack formed, the ice hardened around it, so subsequent cracks formed parallel to the first one, not being able to cross it. Thus a succession of parallel cracks or valleys developed. Over subsequent eons, the icy topography slowly deformed and smoothed out the rough edges of cracks, yielding the bundles of grooves visible today. One would feel more comfortable thinking of grooves as the normal response of ice, rather than rock, to extensional stress if similar patterns existed on other icy worlds. Although Voyager 2's remarkable portrait of Enceladus hints that Ganymedian terrains exist on that small world, most icy satellites of Jupiter and Saturn are without grooves. Perhaps it is just as well to leave the unravelling of such mysteries to the Galileo orbiter mission, which is intended to visit Ganymede and other Jovian satellites in the early 1990s, barring further budget cuts (see Chapter 13).

Of the four Jovian moons discovered by Galileo 3½ centuries ago, I have saved Europa for last. Moving outward from hyperactive Io, through once-active Ganymede, to the ancient, dead world of Callisto, one might expect the second innermost moon, Europa, to be moderately active. We already knew before Voyager that Europa shared properties with both its neighbors, Io and Ganymede. Its bulk composition is that of rock: a bucket of Europa would weigh three times a bucket of water or ice. Yet Europa is mostly covered with fresh ice, even more so than Ganymede.

When Voyager 1 photographed Europa from afar during the March 1979 flyby, the bright second satellite of Jupiter looked much like a well-used ping-pong ball. It appeared perfectly smooth and spherical, without a hint of mountains or canyons. It was featureless, except for some delicate lines or cracks, reminiscent of the fictitious canals of Mars. Voyager scientists eagerly anticipated the closer views to be taken by Voyager 2 on July 9, 1979. Larry Soderblom, deputy leader of the camera team, predicted that the transitional moon Europa "could be the most exciting satellite in the whole Jovian system."

Scientists were disconcerted and perplexed by the later pictures of Europa. Although they were five times sharper than those obtained in March, there were still no mountains or valleys to be seen. Europa is a world the size of our moon, but its tallest hills and ridges are just a few hundred feet high! Furthermore, it is almost as devoid of craters as Io! Only three or four have been found. Either Europa's surface cannot retain a crater once formed, as though it were a liquid ocean, or the craters are buried by geological processes that are as active as those on Earth. A newly formed crater 6 miles across would be erased by the mysterious Europan forces in only 30 million years—less than 1 percent of Europa's age.

Europa's smooth surface is not exactly bland and devoid of detail, however. The earlier hints of "canals" resolved into intricate networks of stripes, crisscrossing the planet. They occur in different shades of white, orange, and brown. Stripes range from 10 kilometers in length to a few that stretch more than halfway around the globe. At least eight different families of stripes have been distinguished by their orientations, widths, lengths, and colors. A few stripes appear to be low scarps or ridges, but most look like they were "painted" onto the flat surface: cracks that have somehow healed.

What exactly is happening on Europa? Even Voyager 2's close-ups were taken from greater distances than our views of the other Galilean worlds. They are not really sharp enough to characterize Europa's diminutive vertical relief, the critical third dimension that is so important to a geologist trying to understand a landscape. Nevertheless, fertile theoretical minds have already been trying to formulate a believable model, or group of hypotheses, to explain Europa.

The first idea was simple enough. Europa, after all, is caught in the same orbital vise grip with Ganymede that keeps Io heated by tidal flexing. Portions of Europa might be expected to have melted just as Io has melted, although less severely because it is farther from Jupiter. Perhaps Europa consists of an Io-like rocky body, reasoned Pat Cassen, surrounded by an ocean of liquid water nearly 100 miles thick and capped by a thin ice crust. Combined radioactive and tidal heating might have kept the ocean mantle from freezing. The postulated thin ice crust would not support tall mountains or deep canyons. Cracks in the ice crust would permit internal water to spill out, rejuvenating the surface. These ideas accounted for all the main traits of Europa —its bulk density, its lack of mountains and craters, and the white, icy, cracked surface.

Unfortunately, it now seems that tidal heating of a Europan ice crust may be less effective than Cassen first thought, so the oceanic mantle might have frozen long ago. Aqueous volcanism would have stopped and it would be difficult to understand why Europa's surface is so young and crater-free.

Two geophysicists at the Jet Propulsion Laboratory, Tony Finnerty and Gary Ransford, started to rethink the Europa problem. Instead of visualizing Europa's water enveloping a dry, rocky core, they imagined Europa was made of certain "wet" or hydrated minerals that have water bound into their crystal structure. As Europa's interior warmed to temperatures exceeding 850 K, the water would have been driven out of the minerals and would have percolated upwards through Europa's interior. Eventually the upper 300 kilometers of Europa's mantle would have been saturated with water so that the crystals could hold no more and all the pore spaces between mantle rocks would have been filled with water, expanding the whole planet. Since water-saturated rocks are very brittle, Finnerty, Ransford, and their colleagues think that the base of the hydrated mantle might have begun to crack.

A crack pierces its way upwards, driven by the pressurized water and steam. Water, plus rocks ripped from the walls of the developing fissure, explode onto the surface of Europa in a violent fountain, according to this scenario. It is somewhat like an aqueous version of the so-called kimberlite eruptions that have occasionally propelled rocks from the upper mantle onto the Earth's surface in South Africa, the Four Corners region of the Southwest, and elsewhere. Kimberlite eruptions leave vertical "pipes" in the ground which are mined for diamonds. But on Europa, the weak, brittle ice-rock crust might split laterally, Finnerty and Ransford think, which would cause the crack to spread across Europa's surface. The observed lengthy bands of fresh rock-contaminated ice are just the surface expression of extensive, thin, vertical sheets (instead of pipes) of the freshly frozen ice-rock materials that forced the cracks apart from below.

It is difficult to visualize the hypothesized aquaeous volcanism on Europa. At the initial site of an eruption, a watery flood spurts onto the surface, flowing miles from the fissure before freezing. How fast does a fissure spread? Perhaps it rips across Europa's landscape at sonic velocities: in a few minutes, the crust is violently rent apart for hundreds or thousands of kilometers. Alternatively, perhaps episodic pipelike eruptions take years or millennia to form the lengthy bands. Or perhaps Finnerty and Ransford are wrong and the stripes are forming in as-yet-unimagined ways.

Scientists still don't understand why Europa's surface appears so young. A hydrated Europa may be more effectively heated by tides than the ice-rock model studied by Pat Cassen. Or perhaps upward percolating water leaches enough salts to prevent a sub-crustal ocean from freezing, even at low temperatures. For now, we must be satisfied with a qualitative portrait of Europa, which stands intermediate between Io and Ganymede both literally and figuratively. Europa's narrow, scalloped ridges, greenish patterned ground, lumpy brownish hills, and confusing families of stripes remain mysteries. Voyager 2's pictures were good enough for mapping the shapes and colors that are unique to Europa but not good enough to reveal the landforms that could provide diagnostic clues about the actual processes of Europan geology. For that purpose, the Galileo orbiter mission is waiting in the wings.

As the Voyagers sailed on toward Saturn, four brand new worlds were left unveiled for scientific scrutiny. Io is the strangest, most relentlessly dynamic solid planet in the solar system.

Europa is the one most devoid of topography. Ganymede's surface was frozen in the midst of developing a water-ice analog of terrestrial continental drift. Saturated with shallow craters, Callisto shows us what our own moon would look like were it made of ice rather than rock. Few geoscientists would deny that the Voyager missions may have added more to our knowledge of extraterrestrial geology in a few short days in March and July 1979 than we had learned in the entire history of planetary exploration dating back to Galileo himself. Each Galilean world was found to be unexpectedly unique. Yet each could begin to be understood in relation to its neighbors as resulting from evolutionary processes that varied with distance from Jupiter. By the end of the 1980s, planetary scientists will have compared the new data about Jupiter's moons with the inner planets and with the smaller Saturnian moons observed by Voyager in 1980 and 1981. A new planetological synthesis will be ready for testing as the Galileo spacecraft approaches its rendezvous with the Jovian system.

10

Saturn Encounter: Resplendent Rings, Exotic Moons

Speeding through space—speeding through heaven and
the stars;
Speeding amid the seven satellites, and the
broad ring . . .
—*Walt Whitman (from "Song of Myself," 1855)*

A golden globe, encircled by a shimmering silver ring: Saturn, the archetypal planet! Poets have sung of rings and painters have portrayed ringed worlds in the heavens. A Saturnian logo topped the *Daily Planet* skyscraper in *Superman*. Recently we learned of rings around Jupiter and Uranus, so ringless worlds may indeed prove the exception rather than the rule. But for three centuries, since Christiaan Huygens recognized Saturn's imperfectly seen "handles" to be a "thin flat ring" surrounding but nowhere touching the planet, Saturn's ring has seemed majestically unique. A standard astronomy textbook of 50 years ago described Saturn this way: "By a good telescope, it is revealed as a delicately banded globe poised within a shining *ring*—a spectacle which, once seen, can never be forgotten. So far as is known, no other body in the universe is like it."

I remember well my own first view of Saturn through a 2-inch refracting telescope my father had just built. It was late in a summer's evening in the early 1950s. The dim but steady yellow star had already dipped behind lofty elms across the street from our yard, so my father pointed the telescope from an open second-floor window, toward the northwest. I knew from books how Saturn *should* appear, but to see the wondrous ring with my own eyes was like magic. As Saturn moved relentlessly toward the treetops, we regularly recentered the cardboard-tubed instrument and I eagerly observed the ringed planet and its two visible moons.

A couple of years later I was mistaught in school that Saturn's rich yellow hue was due to golden sand dunes that covered the planet's surface. Long after learning that Saturn actually is a cloud-bedecked, surfaceless globe of gas, I remained troubled by my misconception of a Saharan Saturn. Many years later, on a still autumn night, I was afforded the best view of that distant world possible from Earth. Within a dome hidden amid Coulter pines on the Los Angeles Basin's northern rim, I paused in my nocturnal routine. Below the venerable Mount Wilson Observatory, Southland folk recuperated from the day's choking inhalations. My assistant turned the ponderous 60-inch telescope toward Saturn. Eventually, the immense, crisp image of the planet floated into my field of view. Amidst a sparkling entourage of seven glittering satellites and encompassed by the crystalline rings, Saturn's oblate globe was tinted with a dozen hues—pastel pinks, creams, and yellows tinged with gold. Stilled by the atmospheric inversion, the planet appeared ten times sharper to me than it had to Huygens three centuries earlier, during the dawn of the Age of Telescopes.

For three decades since I first observed it as a boy, Saturn has revolved before the constellations of the zodiac. Now it has fully circled the sun, completed its cycle of seasons, and has begun another of its protracted "years." Similarly, Saturn has revolved 11½ times since the first astronomical telescope was turned toward it. Ninety Saturnian years have elapsed since the planet was first recorded in the dawn of human history; 150 million times before that Saturn orbited the sun unwatched. Suddenly in mid-November 1980, Saturn's secrets were unveiled when the multi-instrumented Voyager 1 spacecraft sped close to the planet. Voyager broadcast a stream of digital data to Earth that was reconstructed into spectacular photographs a thousand times sharper than our best views from Earth.

On those dramatic days and nights, I was privileged to sit in the inner sanctum of the Voyager Imaging Team at the Jet Propulsion Laboratory in Pasadena, California, only 5 miles below Mount Wilson where I had observed Saturn from afar. Here in a crowded suite of rooms on the fourth floor of JPL Building 264, 18,000 pictures of Saturn were received and studied by the men and women who had designed the spacecraft's camera and had planned the picture-taking sequences. The feeble signals from Voyager's 20-watt transmitter had traversed the solar system for an hour and a half before being collected by 210-foot antenna-dishes in Australia, Spain, and the Mojave Desert. The signals were transmitted instantaneously from those Deep Space Network stations to Mission Control at JPL, translated into TV pictures, and flashed on the screens I was watching.

Enthusiastic scientists crowded around two computer consoles, playing with recently received pictures: magnifying them, changing brightnesses, improving contrasts. Once a rendition was perfected, the push of a button transformed the flickering TV image into a flimsy, hard-copy print. Scientists showed some prints to each other and taped others to the wall. When critical pictures were due, the men and women expectantly watched the suspended TV monitors for brand-new snapshots from Saturn, which appeared every 2 or 3 minutes.

The frenetic activity lasted around the clock during the few days closest to the November 12 encounter. Satellite experts caucused in one room while ring-specialists mulled over mystifying data in another. An NBC *Today* show crew came through the team area after midnight to videotape some exhausted scientists viewing new pictures of satellite Rhea's crater-scarred surface.

In nearby Von Karman Auditorium, typewriters in the press

area clattered throughout the night and floodlights glared as reporters fed live spots to news programs already beginning on the East Coast. Carl Sagan and other scientist-spokesmen stood before a glittering, life-sized model of the Voyager spacecraft, stoically trying to look engaging on *Good Morning America* before getting back to bed. As misty sunlight began filtering through the eucalyptus trees, the print-media reporters crowded back, grabbing up mimeographed agendas for the daily morning press briefing. Lucky ones known to the stingy picture purveyor were handed packets of glossy panoramas of Saturn. Ducking cameras, dodging celebrities, and stepping over a maze of cables, journalists vied for seats with space groupies, who had managed to wangle press badges to report for a hometown newssheet or radio station.

Anointed Voyager experts hurried to get slides of last night's data before ascending the stage. They hastily conferred about whether their previous evening's brainstorming ideas, lubricated by celebratory champagne, were solid enough to be fed to the gathering throng. A stern JPL press officer quieted the press corps and lectured them about violating journalistic etiquette, such as interviewing busy scientists without his permission. Excluded from the press area, most JPL scientists learned of their colleagues' latest findings from the live broadcast of the press conference shown on monitors around the laboratory. The briefing began. Statements were read, questions answered: words that soon were indelibly part of the public record around the world. Some were memorable insights, others were sloppily reported as heralding the overthrow of physical laws. Still others were premature interpretations that would eventually succumb to sober scientific analysis.

For many scientists and science journalists alike, this was an especially memorable and poignant planetary encounter. There had been many before; there might even be many again. But in a solar system of just nine planets, there could be only a few *first* encounters. And what a world, what a system, Saturn was to explore for the first time! Although Pioneer 11 had scouted Saturn a year earlier, as the Norsemen had scouted America, Voyager was the Columbian expedition of discovery for the outer planets. For many Voyager scientists, including Bradford Smith, leader of the camera team, the heady plunge past Saturn was the pinnacle of Voyager's flight. Said Smith, a veteran of Voyager's revolutionary Jupiter encounters and many previous missions, "I cannot recall being in such a state of euphoria for any planetary encounter."

Time seems distorted as a spacecraft approaches its planetary target. When you walk toward a skyscraper or drive toward a faraway mountain range, your goal long remains discouragingly distant, growing imperceptibly larger. As you near your destination, the mountains suddenly envelop you or the building looms above you. For months Saturn seemed as sharp to Voyager as in a telescope from Earth, yet from day to day its familiar aspect hardly changed. Finally Saturn grew larger and its rings were resolved as clearly as in Pioneer's portrait, laboriously reconstructed line-by-line the previous year. During October 1980, the pace of discoveries quickened each week. Spinning spokes appeared in the rings. The three broad rings resolved into bands, separated by gaps. In quick succession, ringlets appeared in the gaps and bands, then divisions within the ringlets. During the final day, the rings expanded at a dizzying rate, dividing into a myriad of concentric circles, nested within one another. Voyager plunged brazenly through the ring-plane, then its camera glanced upward at the whirling annulation's translucent undersides. Suddenly flung up and away by Saturn's powerful gravity, Voyager looked back on a remarkable, receding vision of the ringed giant: a strangely glowing halo bifurcated by the columnar shadow of a crescentic Saturn. That haunting farewell view that would last for years, dwindling ever more slowly in Voyager's lenses as the spacecraft glided out into the blackness of space.

Nearly all Voyager's discoveries were compressed into a day and a half. The first body to be approached closely was not Saturn, but its largest moon. Although Titan must defer to its splendid parent, it is appropriately named, being a world larger than the planets Mercury and Pluto. It is the only planetary satellite to possess much of an atmosphere, first detected spectroscopically by Gerard Kuiper during World War II. Theoretical analysis of later telescopic spectra suggested that Titan is a murky, smog-ridden place. Optimists such as Carl Sagan hoped Voyager's cameras might pierce the haze and record Titan's mysterious surface through a gap in underlying clouds.

No such luck. On Tuesday, November 11, Titan's featureless facade loomed ever larger in Voyager's lenses. A thin layer of bluish haze was profiled 100 kilometers beyond the edge—the horizon—of the bland, orange disk. To the north, the layer thickened and merged with a gloomy, inpenetrable, brownish-red hood which shrouded Titan's polar region. During the evening, Voyager passed by Titan, concluding its sequence of disappointing closeup pictures. Titan had turned out to be just a "fuzzy ball covered with brown smog," as one geologist put it. At 10:45 P.M., Bradford Smith looked away from the monitor and said to no one in particular, "Well, so much for Titan."

It was a reaction shared by many of the camera-team scientists, but actually the probing of Titan's atmosphere was just commencing. Nine minutes later, the antenna in Madrid, Spain, homed in on Voyager's radio beacon as the spacecraft's trajectory carried it behind Titan. Slight changes in frequency or tone of the fading signal would probe the temperature and pressure through the thick smoggy atmosphere down, scientists hoped, to Titan's ground, ocean, or whatever. It was touch-and-go for a while as an unexpected morning thundershower over Madrid threatened to distort a transmission that had traveled unimpeded by Titanian smog and a billion and a half kilometers of interplanetary space.

Overnight, experimenters did their "instant science." Spokesmen emerged the next morning to report that Titan's atmosphere was hundreds of kilometers thick and perhaps five or ten times as massive as Earth's. It is the only atmosphere besides our own composed mainly of nitrogen. "Titan may be considered a terrestrial planet in a deep-freeze," said one scientist who suggested that Titan might even have lakes of liquid nitrogen and high-altitude clouds of nitrogen droplets.

Later analysis modified the picture. Titan's nitrogen air is cold and thick, at a barometric pressure half again higher than on

Earth, but the climate is not quite cold enough for nitrogen to liquefy. Instead, the monotonously frigid Titanian weather is controlled by methane (CH_4), or "marsh gas." Although methane, the chief component of natural gas, makes up only a small percentage of Titan's air, it coexists simultaneously as gas, liquid, and solid at Titan's surface temperature of 92 K above absolute zero (−295°F). Titan's methane thus might act the way water does on Earth near 32°F. The view on Titan's surface at this "triple-point" of methane would be richly varied but weird. A Titanian inhabitant might peer across an arctic sludge-scape, dully illuminated by a blood-orange sky. The "rocks," made of methane and ammonia ices, would be mostly buried beneath a gooey layer of soil, accumulated over aeons from the continual settling of tiny dust specks of tarry smog, formed near the top of the atmosphere. Titan's soupy lakes and oceans of liquid methane would often be hidden by choking mists of methane ice-fog. High overhead, through the slightly hazy air, the Titanian resident would see reddened clouds of liquid methane droplets.

Ascending 200 kilometers to the smog blanket's top, the Titanian might marvel at bloated Saturn and its edge-on rings materializing from obscurity. Here in a rarefied climate like that atop Olympus Mons on Mars, the distant sun manufactures photochemical smog, using Titan's natural methane instead of automobile exhaust as the raw material. As the smog-dust coagulates and precipitates, fresh emanations of oceanic methane diffuse upward, only to be fragmented by solar ultraviolet rays and energetic magnetospheric atomic particles. Propane, ethylene, and acetylene are produced, then are polymerized into more complex organic molecules. The byproduct, hydrogen, floats up and escapes to space, forming a gigantic, flattened hydrogen donut enveloping Titan's orbit and girdling Saturn.

Here in Titan's thermosphere, just below the interface with outer space, seasons are especially apparent. Each one lasts 7 Earth years, and is marked by different temperatures, colors, and heights of the smog hazes. Due to the energy reservoir of Titan's massive atmosphere, seasons change reluctantly. The warmest temperatures occur in autumn, the coldest in spring. Such laggard seasonal changes fooled Earthbound astronomers who thought that changes over the years in the brightness of Titan and other planets reflected slight variations in the "solar constant" of the sun.

Voyager 1 didn't wait for these follow-up musings about Titan as it sped to the next moon on November 11, 1980. The following morning, it closed on Tethys, a smallish sphere of solid water-ice about as big as the largest asteroid. Of Saturn's nine "classical" satellites, Tethys is third out from the rings. Once only a starlike point of light, Tethys was revealed as a cratered world, split by an enormous branching chasm, twice as deep as the Grand Canyon and several times as long. Voyager 2 would later reveal the chasm to extend around three-quarters of the girth of Tethys. Perhaps the great canyon reflects the splitting apart of Tethys's crust when the interior of the once watery globe froze solid.

Peering ahead to its next target, Voyager captured an uncanny portrait of tiny Mimas looking just like the Death Star battle station in *Star Wars.* Mimas, less than 400 kilometers (250 miles) in diameter, is gouged by an enormous crater a third the size of the satellite itself! Protruding from the crater's floor is a huge mountain, as tall as Mount McKinley. Voyager scientists speculated that the comparatively recent cratering impact might have nearly converted the small icy world into fragments forming an immense new ring. Obviously Mimas was not destroyed. But some scientists think that Mimas may be the remnant of a larger moon that was nearly destroyed by an even greater impact eons ago.

During the afternoon of November 12, while Voyager was underneath the rings, it was also nearest Mimas, which orbits Saturn once a day just 50,000 kilometers beyond the brighter rings. The best views of Mimas, which are of the opposite side from the giant crater, show it saturated by deep crater pits. Nowhere else in the solar system have we seen a surface so densely pockmarked by small craters. Perhaps Mimas once encountered boulders from a now-vanished ring. Ice-boulders ejected from the huge Mimian crater, or by earlier impacts, may have orbited

Saturn and reimpacted on Mimas. In addition, Saturn's moons were all struck by comets, and Saturn's gravity would have concentrated cometary impacts onto nearby Mimas, contributing to its battered terrain. Mimas also has lengthy gorges on its surface, perhaps due to once-active subterranean forces within the now-dead little world.

Later in the evening, Voyager sped on past Dione and Rhea, two larger, more distant members of Saturn's satellite system. At first they seemed denser, or heavier, for their size than Tethys or Mimas, suggesting that they were composed of rock in addition to ice like that visible on their surfaces. Nongeologists became a bit bored to find that Dione and Rhea also had canyons and crater-scarred surfaces, but specialists marveled at the diversity of craters. Many battering impacts on Rhea formed polygonal craters, some seeming more nearly square than round. Sparsely cratered plains on Dione hinted that it was once warm enough in places for subsurface ice to become slushy. Conceivably Dione's strange control of Saturn's radio noise implies that it is still outgassing or is otherwise active. The strangest-looking part of Dione, as viewed from afar, is the middle of its trailing side, which has been largely protected from erosion by the blizzard of meteoroids encountered by the leading side. However, the backside could not be photographed at close range by either Voyager 1 or 2. So if its well-preserved, small-scale features hold clues to Dione's ancient past, those mysteries must await a follow-up Saturn mission.

Tantalizing pictures of more distant moons indicated they also had unique personalities. Several theorists speculated that Enceladus, apparently devoid of giant craters, might have only a thin shell of surface ice. Its watery interior, possibly spiced by ammonia "antifreeze," may be tidally warmed due to a resonant lock with Dione. Impacts might occasionally puncture the ice. Geysers fountaining water into space could then replenish the tenuous E-ring of Saturn with its sparkling ice crystals. The hypothesis gained some support after Voyager 2 zeroed in on Enceladus in August 1981 and revealed some of its surface to be the youngest terrain yet seen in the Saturnian system.

Differences between crater populations on Mimas, Tethys,

Dione, and Rhea probably reflect different numbers of Saturnian ring particles and ejected crater fragments impacting on each moon. These populations could also have been affected by the different times when each moons' internal geological activity finally froze into quiescence. If we could disentangle these confusing elements, we might be able to decipher the record of cometary bombardment on these bodies. Unlike the inner planets or even Jupiter's satellites—which are cratered partly or totally by asteroids or dead comet cores—the Saturnian moons intercept only pristine comets. A prime goal of research on Saturnian cratering will be to learn more about these denizens of the frigid outer solar system. Comets may be remnant building blocks of the outer planets, but the few that stray into the inner solar system evaporate away most of their icy bulk in that comparatively ovenlike environment. Those that eventually crash into planets are corpses of their former selves, their essence having streamed away in the flashy tails of gas and dust that have awed mankind for millenia. The best hope for learning about the numbers and sizes of fresh comets over the history of the solar system may be to study their scars on Saturn's moons.

In the late 1960s, the French astronomer Audouin Dollfus observed a tiny satellite just beyond the visible rings. He named it Janus. Instead of being two-faced, Janus turned out a decade later to be two separate satellites, sharing the same orbit. Voyager's camera was aimed at each of them and showed both to be oblong and misshapen—possibly two halves of a demolished satellite. A dark band unexpectedly swept across the surface of one of them during the picture-taking sequence. It later proved to have been the shadow of the slender F-ring.

During the years and months before the Saturn encounter, ground-based astronomers raced to discover new Saturnian moons before Voyager beat them to it. The immediate outcome was a tie: Astronomers 3, Voyager 3. (Additional moonlets have been found since.) One Earth-based discovery was of a tiny satellite in Dione's orbit. The Dione and Janus coorbital pairs were the first cases known of solar system satellites sharing orbits. The three new satellites discovered by Voyager are each about 100 kilometers in size, located just barely outside the major rings.

They turned out to have intimate relationships with the rings that blur the distinction between satellites and rings. What is a narrow ring if not a multitude of tiny moonlets sharing the same orbit?

If the geologists and astronomers were overwhelmed by Saturn's multiplying microcosm of moons, everyone was dazzled and stunned by the rings. The sheer expanse of the ring system, from beyond the orbit of Enceladus all the way down, perhaps, to Saturn's atmosphere, had to be experienced during encounter to be fully appreciated. For hours Voyager sailed beneath the canopy of rings at 100,000 kilometers per hour, snapping exquisitely detailed pictures. Voyager scientists taped pictures spanning the major rings to a wall; they stretched nearly from floor to ceiling. It took three ringlet-covered pictures laid end-to-end just to cross a supposed ring gap that can be glimpsed with Earth-based telescopes as only a single darkish line only when atmospheric conditions are good.

The intricate structure of innumerable ringlets was totally unanticipated. The bewildering complexity of rings within rings within rings . . . intertwined rings shepherded by tiny satellites . . . shadowy spokes on approach that became iridescent on departure . . . gaps and divisions, some seemingly randomly spaced, others in an orderly geometric progression . . . rings of snow boulders, others of fine ice dust . . . eccentric rings, tilted rings, and clumpy rings. The rings were not, after all, the simple, planar expanses astronomers had expected. The unusual phenomena and remarkable intricacies cried out for scientific explanation.

Astronomers are loathe to make hypotheses about nature that are any more complicated than minimally necessary to explain observations. Saturn's wonderful rings, as observed from a distance, were never hard to understand. Or so astronomers thought! At first, they were thought to be solid annuli, like cardboard disks. Then a century ago James Clerk Maxwell presented theoretical proof that they must be composed of numerous independently orbiting bodies. Subsequent observations with a spectroscope, employing the so-called Doppler shift principle, verified

Maxwell's conclusion. The inner rings were approaching (or receding) faster than the outer rings, and they all were moving at velocities appropriate for particles in Keplerian orbits. (Kepler's Third Law states that the square of a particle's period of revolution is proportional to the cube of its distance from the central body, in this case Saturn.)

A few problems remained: to determine the size and composition of the particles and to explain several gaps between the rings. The gaps seemed easy. The chief one, called Cassini's Division, is located where a ring particle at its inner edge would orbit Saturn exactly twice for every orbit of the nearby satellite Mimas. Thus Mimas's gravity would periodically tug at such a particle, dislodging it from its orbit and clearing the gap. Only the great breadth of the Division was a bit puzzling, but theorists didn't worry much about it. They needn't have wasted their time. As it turned out, the "Division" is mostly filled with ringlets and it is fortunate that the original plan to fly Pioneer through its middle was changed.

Until a decade ago when new spectra were obtained, it was a mystery what the ring particles were made of. After some initial confusion, astronomers agreed that the deep absorption bands were uniquely due to water ice or water frost. Later, radar observations proved that the frost is not just superficial but that the bulk material is probably water ice, too.

The final question about the rings was how big and how numerous the particles are. Clearly the rings are more than an occasional snowflake here and there, for they are so bright that there can hardly be much black space showing through. On the other hand, when Saturn occasionally passed in front of a bright star, astronomers could see the star dimly through some of the rings, ruling out a blizzard of particles. So the rings are a flurry of particles. Analysis of the rings' reflected light suggested the particles were smaller than several inches in size. Yet they proved to be extraordinarily reflective to radar waves, not transparent as expected for particles much smaller than the wavelength of the radar waves. How could the particles be simultaneously large and small? Mystified astronomers for a while resisted the simple answer: there are particles of *both* sizes . . . and probably other sizes as well! Larger bodies must, after all, exist to replenish the snowball-sized ones, which are rapidly eroded and destroyed by meteoroids and smaller impacting particles. Such potential ring-moonlets would be invisible from Earth if they were smaller than 100 kilometers in size, even if they accounted for 99.9 percent of the mass in the rings.

As Voyager 1 approached Saturn, astronomers' best models for the rings resembled the uniform, translucent, expansive ribbons so often air-brushed in the past by space artists trying to portray Saturn. The supposed swarm of football-sized snowballs would never be resolved by Voyager's camera, which could not photograph ring structure smaller than a few kilometers. The scattering of Voyager's radio beam through the rings as the spacecraft departed Saturn suggested that many ring particles are as big as a house. As it turned out, still larger moonlets imbedded within the rings were not easily detected by Voyager 1. But they were quickly hypothesized to exist as one of several possible explanations for the remarkable ringlet formations.

Theorists turned to Saturn's magnetosphere—the volume of space in which the motions of charged atoms and molecules are controlled by the magnetic field generated within Saturn—to help explain the B-ring's wheeling spokes, which are probably related to the lightninglike static heard by Voyager's radio receivers. (The spokes are ever-changing radial dark and bright patches that cross Saturn's broadest ring, shown in Illustration 31. Voyager 2's camera zoomed in on several spokes racing around the planet, resulting in a dizzying movie, which scientists aptly dubbed "the Voyager 500.") As theorists reconsidered the processes of mutual collision, fragmentation, and charge separation among icy ring particles, they began to appreciate ways in which a summer hailstorm might have a few things in common with Saturn's icy rings. Magnetospheric effects also were examined to help explain the apparent braiding, or three-dimensional intertwining, of the separate strands that make up the F-ring. Celestial mechanics experts reexamined whether periodic gravity effects of the larger moons could generate the lopsided rings, as well as some of the more regularly spaced gaps.

Could the intricate structure of the ringlets and gaps be telling us about the still smaller but fundamental structure of the rings —the moonlets and particles too small to be photographed? Probably so. The relationship between the three new satellites discovered by Voyager and some nearby rings is a clue. For instance, an elongated moonlet 20 kilometers wide and 40 kilometers long orbits just 800 kilometers beyond the classical rings' outer edge. Any ring particles that stray toward the satellite would be deflected back or swept up by the moon, but could not cross its orbit. Thus the moonlet defines the outer boundary of the rings. Perhaps it is even a source of ring particles, chipped off by meteoroid impacts.

The interactions between the narrow F-ring and its shepherding pair of moons is especially striking. The F-ring is bounded on the inside by one small moon and on the outside by another, each about 1,000 kilometers from the ring. When the satellites were first discovered, they appeared to be a textbook example of the shepherding process conceived theoretically 2 years earlier to explain the then newly discovered narrow rings of Uranus. Ring particles approaching the more slowly orbiting outer moon are first pulled outward toward the moon. But as they pass on ahead, they are dragged even more strongly by the gravity of the lagging moon. There is a net loss of energy, so they move inward to an orbit farther from the moon and closer to Saturn than the original orbit. The inner moon exerts exactly the opposite effect. So acting together, the two moons confine F-ring particles to a narrow zone between them, thus forming a ringlet. Or so scientists were thinking during the final days before encounter. As Voyager flew close, it discovered three separate strands of the F-ring to have an inexplicable braided and knotty structure. While the elegant simplicity of theory was thus dealt a temporary setback, analogous processes of confinement could occur throughout the ring system, involving moonlets perhaps 100 meters to 10 kilometers in size. That may be how some ringlets are formed.

On the evening of encounter day, I found Fred Franklin of the Smithsonian Astrophysical Observatory leaning back in a chair, staring at the wall. Nearby on an interactive computer monitor was a close-up of the rings. Fred was an old hand at Saturn's rings. He had been publishing papers about them back when he was my advisor while I was an undergraduate in college. Now, in a few days, his patiently acquired knowledge about the rings had become suddenly obsolete. The revealed truth was now flickering before him on the screen. I asked him what he had learned. He stared at the monitor and sighed, shaking his head slightly in bewilderment. "Oh, something about myself," he finally replied. I waited for him to elaborate. "There are gaps in Saturn's rings and there are gaps in my knowledge."

Months later, the ring data were finally digested, if not understood: all the pictures on approach, looking back, looking up at the dark side, pictures of Saturn through the rings, pictures of the ring shadow on Saturn, temperature profiles, radio probings of the rings like those of Titan's atmosphere, and more.

The next summer, Voyager 2 sailed past Saturn with its computers reprogrammed to execute picture-taking sequences that would follow up on the discoveries of Voyager 1. Exciting movies were constructed showing the B-ring spokes reeling around Saturn as if on a gigantic merry-go-round. The ringlets were resolved into even more exquisite detail and were found to be changing position, like the incessant motion of ripples on a pond. But the pictures targeted on the F-ring at first showed not a hint of braiding; rather, its multiple strands were perfectly colinear, gently mocking the scientists' hastily contrived explanations for its ephemeral appearance the previous year. One instrument on Voyager 2 measured the fluctuating brightness of a distant star as the rings passed between it and the star. The resulting trace revealed that the intricate multiplicity of ringlets seems to extend right down to the scale of the boulders that compose the rings!

But the ringlets still mystify the Voyager scientists. A year after Voyager 1's historic encounter, and months after Voyager 2's follow-up, researchers were still searching the pictures in vain for the tiny moonlets thought to be clearing the gaps and confining the ringlets. It will be a formidable, long-term challenge for analysts and theorists to decode the message in the multiplicity of ringlets. Perhaps one day we will understand the mélange of infinitesimal to tiny to not-so-tiny bodies that compose the rings of Saturn as well as the forces that govern their striking behavior. One thing is clear: nature has proved again to be unexpectedly fascinating. It is one more lesson that exploration of our world, our cosmos, will never fail to be richly rewarding.

11

Mars: Changing Perspectives

"It's just incredible to see that Mars is really there!"
—Thomas A. Mutch, 1976

Seven years to the day after a man first walked on the moon, Tim Mutch was savoring the most rewarding moment of his career. Of course Mars was there. Moreover, the Viking spacecraft camera was working! Tim Mutch had been responsible for the team of scientists and engineers that had designed, built, and tested the camera that was at this moment sitting on the orange plains of Mars transmitting to Earth the first close-up of a Martian desert. He was sharing his personal triumph with millions of viewers, watching over his shoulder via network television, as the sliver of a picture gradually expanded into a full-size portrait of the ground beneath Viking's footpad.

The second picture began to develop, left to right—an eye-level panorama of a boulder-strewn scene. Viking scientists were thrilled as they watched their TV monitors at the Jet Propulsion Laboratory. Yet something was wrong. The sky was bright; indeed, as later color pictures revealed, it was bright ocher. Where was the blue-black sky that scientists had expected?

This was to be only the first of many surprises for Viking scientists as the spacecraft's sojourn on Mars lengthened from hours to days, then months. Elated when the three biology experiments seemed at first to indicate abundant life on the Red Planet, researchers were dumbfounded when the chemistry experiment failed to prove the existence of any organic molecules on Mars. The whole issue of Martian life was cast into a state of ambiguity. Meanwhile, overhead, the first Viking Orbiter, later joined by a second, determined that Mars's presumed permanent dry-ice polar caps are in fact composed mostly of water-ice. Later the Orbiter cameras photographed mysterious grooves on the largest of the planet's two moonlets.

The first months of the Viking mission epitomized our ever-changing ideas about Earth's ruddy neighbor in space. Ancient associations of Mars with a warlike deity, overprinted by Percival Lowell's turn-of-the-century theories of a canal-building Martian civilization, culminated before World War II in Orson Welles's terrifyingly real radio drama about an invasion from Mars, which few listeners then doubted was teeming with life.

On July 30, 1965, there was again news from Mars. The 18-point headline on the front page of the *New York Times* reported scientists' pessimistic conclusions from the first Space Age reconnaissance of the planet:

MARINER 4'S FINAL PHOTOS
DEPICT A MOONLIKE MARS

Signals radioed across millions of miles from the 575-pound spacecraft had been reconstructed by NASA computers into fifteen closeup pictures of a crater-scarred landscape. As the *Times* went on to report,

> A heavy, perhaps fatal, blow was delivered today to the possibility that there is or once was life on Mars. The

Mariner 4 photographs taken of the planet at close range July 14 show a crater-pocked landscape lacking any sign that there has been water erosion there.

Public fascination with Mars died a death as bleak as the "lifeless" landscape shown on the wirephotos, and the nation's concern was directed inward to a divisive war and social upheaval. A few years later a vice-president suggested sending men to Mars by the 1990s; a bubble of public indignation arose over the outrageous cost of such folly, then subsided again.

Meanwhile, oblivious to the death sentence they had imposed, scientists struggled to unravel the secrets of the Red Planet. Bigger and better spacecraft were launched toward the "moonlike" world. The fifteen close-ups became dozens, then thousands. The surface of Mars was found to be covered not only with craters but also with volcanoes, canyons, and hundreds of dry river valleys, suggesting that Mars had had a more clement climate in its past, after all. Work proceeded on developing an unmanned mission to land on the Red Planet in order to learn about Martian biology.

In America's bicentennial year, the first Viking lander stuck out its mechanical arm and gobbled up the dusty orange soil of the Chryse plains in the hopes that Martian microorganisms would chemically reveal their presence. After studying the data telemetered back to Earth, scientists once again issued their preliminary findings to inquisitive reporters. The data at first seemed ambiguous because Mars is not the precise laboratory environment imagined by the designers of Viking's experiments. Nor would the erstwhile Martian organisms fit precisely the models of recognizable life forms that biologists had constructed from their Earth-biased perspective. Finally, hopes that Martian microbes are alive today faded. Yet the possibility that life evolved on Mars in the past seemed greater than ever as other Viking instruments reinforced ideas that the Red Planet once had a thick atmosphere and a watery environment.

While the Soviet Union focused on Venus during the 1970s, America's exploratory efforts concentrated on Mars. Discounting the Apollo moon program as a piggyback ride on a technological extravaganza, the Viking Mars mission was the most elaborate and costly program of scientific exploration in our history. It was also gutsy, with high scientific stakes: to land blindly on the unpredictable surface of another world and search for alien life. Sheer exploration was the heart of Viking. After all, it was biologists and geologists who pushed for Viking and made it work. And of all the disciplines of planetary science, the tradition of exploration runs deepest in biology and geology. Tim Mutch, the consummate planetary geologist who became NASA's chief scientist as Viking's mission drew to a close, was exploring the high Himalayas when he fell to his death in 1980. Tim was not the only Martian explorer to lose his life on a mountain expedition: biologist Wolf Vishniac died surveying a Mars-like habitat in Antarctica.

Scientists compare observed facts with theory or preconception. What is observed may be examined from a distance or manipulated in a hands-on experiment. Astronomers observe, chemists experiment. Exploration is transitional between observation and experiment, with an intimate, physical interaction between explorer and environment. Objects are collected and manipulated and the state of nature is observed interactively, at close range. After a succession of observational flybys and an orbiter, Viking was a shift from reconnaissance to exploration of Mars. A pair of Viking orbiters scrutinized Mars and took highly magnified pictures of exquisite sharpness, for the first time revealing landforms on a human scale (the best pictures would have shown a truck on Mars). The scientists who manipulated the remote-controlled arms, instruments, and cameras on the landers could dig ditches, test soils, and generally interact with observable things on the patches of Martian ground surrounding the two stations. Viking scientists—geologists in particular—were frustrated, however, by the fixity of the two landers. They knew of much more interesting places on Mars that had been avoided in selecting the safest possible landing sites. With exploratory fervor, they longed to look behind the Volkswagen-size rock, dubbed "Big Joe," and to look over the distant ridges. They wished Viking 1 had wheels to transport it into the mouth of the nearby dry river-valley. Lewis and Clark would have been poor explorers had they remained stuck on a fraction of an acre of North Dakota.

Even while the Viking mission was in full swing, Tim Mutch and other Viking scientists worked to initiate a follow-on Mars rover mission. Concepts ranged from flying a spare Viking with wheels to a much more elaborate rover vehicle capable of exploring far and wide across the planet. Meanwhile other researchers, whose scientific appetites had been whetted by analyzing moon rocks in their own laboratories, pressed to move Martian science straight to the experimental stage. They advocated the return to Earth of Martian rocks and soils for intensive biological and chemical analysis as the next logical step for the American Mars program. While the two ambitious factions quarreled on scientific advisory committees, government bureaucrats and politicians lowered their sights. In fact, they lost sight of the Red Planet altogether. No credible follow-on Mars mission is now being planned in the United States, although the European Space Agency may launch a new Mars orbiter with the new Ariane rocket. Still, the legacy of observed "facts" collected during 4 years by two Viking orbiters and two landers will keep some scientists—the astronomers and photogeologists who feel comfortable with remote-sensing techniques—busy for years to come. Experimentalists and Martian explorers are not so lucky. Evidently they will not be able—soon or in their lifetimes—to satiate their curiosity, aroused by the tentative, imperfect exploratory steps taken by Viking.

As late as the early 1960s astronomers thought Mars might be quite conducive to life. Any doubts they had, sprang from their narrow conception of the tolerance of any Martian life forms rather than from the probability of any especially threatening traits of the Martian environment. (Many biologists, more familiar than astronomers with the remarkable diversity of life, were still more optimistic about the prospects for Martian life.) Mars seemed a pleasant world. Its air was a bit thin by our standards and it got chilly at night, but there were fluffy white clouds, there were seasons like our own, and the Martian day was just 37 minutes longer than Earth's. Polar snow caps melted in the spring and dust storms occasionally blew across the planet, just as in Oklahoma. Flashes of light on Mars, witnessed by Japanese astronomers, were thought to be volcanoes, or maybe even the sun reflected from Martian lakes. Indeed there were signs of life itself. Colors of certain regions changed with the seasons, sometimes becoming bright green. And after a dusting from a passing storm, green patches sometimes reemerged, just like plants poking up through the dust to reach the sunlight. Most conclusive of all, some astronomers thought, was the spectrum of those patches that William Sinton, now of the University of Hawaii, interpreted as suggesting something like chlorophyll on Mars.

Some prescient theorists had other notions. As long ago as 1951 Ernst Öpik had predicted that "the surface of Mars should be covered with hundreds of thousands of meteor craters exceeding in size the Arizona [Meteor] Crater." And a few years later Dean McLaughlin propounded a theory to explain the changing dark patches in terms of volcanic ash being blown about by Martian winds. Öpik's sound logic was ignored and McLaughlin's theory was "disproven." We now know that much of Mars is heavily cratered and that the dark markings are caused by windblown dust, although that dust is probably not chiefly ash from the huge volcanoes that certainly are there.

A few years later Sinton realized that his spectrum was of a whiff of gas in the Earth's atmosphere rather than of plants on Mars. Still, as Mariner 4 approached Mars in summer 1965 most astronomers believed there might be lowly plants on Mars, such as the lichens that encrust desert rocks on Earth.

The men in charge of interpreting Mariner 4's fifteen photographs were not planetary astronomers. One was a physicist, two were geologists (all professors at the California Institute of Technology), for the interpretation of aerial photographs of landforms is geologists' work, not the specialty of astronomers. What the pictures showed was shocking to the three, who had no previous reason for faulting the astronomers' interpretations. Ernst Öpik wasn't surprised, but almost everyone else agreed that Mars was a very different world from what had been imagined.

The best pictures from Mariner 4 showed a bleak, cratered surface. Of course, no photograph could have revealed life itself, unless Martians were several miles across! But the pictures might have shown river deltas, mountain chains, and other evidence of a young, active world like our own. They might even have shown canals built by a Martian civilization.

Instead, our own dead moon, whose surface was then believed —correctly—to be between 2 and 5 billion years old, was the closest likeness we had yet seen to the Mars of Mariner 4. So the Caltech researchers concluded that "the visible surface is extremely old and that neither a dense atmosphere nor oceans have been present on the planet since the cratered surface was formed." Either an ocean or an atmosphere might well have eroded any preexisting large craters, just as on Earth. But were the craters really ancient? That could be determined, roughly, by the technique of crater counting, and three separate groups of scientists immediately challenged the Caltech interpretation of crater ages.

Since they were not astronomers and had no stake in earlier views of Mars, the Caltech professors and other geologists who began studying Mars did little to sort out the significance of the changing colors, the flashes of light, and other earlier evidence of a more hospitable planet. To this day it is not really known which earlier observations were valid and which were illusions. Perhaps they hint at further surprises yet to come. Most astronomers have now gone on to study other bodies in the solar system, leaving Mars to the geologists.

The exploration of Mars was chiefly guided by one of the original Mariner scientists—Caltech geologist Bruce Murray, now the director of the Jet Propulsion Laboratory, which runs planetary space missions for NASA. He, more than anyone else, established the strategy of photographic reconnaissance of the planets. In the early 1960s, with NASA already contemplating a Mars lander project (more ambitious and called Voyager then), Murray argued that the meager pre-Voyager reconnaissance program then planned was totally inadequate. We would be landing on Mars blind and ignorant.

Suppose, for example, that Venusian crocodiles wanted to

search for Earthling relatives. Their approach should *not* be to fly a Mariner 4-type spacecraft to take fifteen pictures of Earth (probably showing ocean) and then drop an expensive lander onto Earth bearing crocodile goodies. Even if their lander didn't drop in the drink, it would be unlikely to land anywhere near a crocodile. A better strategy would be to send a series of reconnaissance spacecraft: the first ones would photograph Earth from distant orbit, revealing the oceans, continents, and mountains; then more sophisticated cameras would zoom in on those border regions between water and land, searching for telltale signs of swamps. They would confidently drop a costly lander only when they had first found a choice swamp.

Murray realized that the then-planned reconnaissance of Mars would have yielded only 1/10,000, or at most 1/1,000, of the information necessary to land a life detector intelligently. Under his leadership Mariners 6, 7, and 9 ultimately made up much, but not all, of the deficit in information. NASA still skipped over the highest magnification zoom-in phase of exploration. That is what caused so much consternation and delay in landing the first Viking on Mars, originally scheduled to coincide with bicentennial celebrations on July 4, 1976. Belatedly realizing that Viking could have impaled itself on never-seen boulders when it settled onto the Chryse plain, project scientists postponed the landing again and again as they desperately sought, from really inadequate data, the safest landing site. Luckily, Bruce Murray's "Martian horror story" of 1964 turned into the astounding success of the Viking mission.

Perhaps the biggest jump in our knowledge of Mars occurred in 1971, when Mariner 9 went into orbit and showed that the crater-scarred landscape of 1965 was not at all typical and that there might be places where Martian life thrived, after all. With hindsight we can now recognize in the pictures from Mariners 4, 6, and 7 telltale signs of unmoonlike geology that we only really noticed once Mariner 9 had photographed all of Mars. The first half-dozen pictures from Mariner 4 showed few craters on a flat and featureless terrain. We now know there are wide craterless plains in that area. But in 1965 the lack of craters was blamed on the great distance from which the pictures were taken and on the poor lighting angle. (Any amateur photographer knows one never gets good pictures aiming directly away from the sun, as was true for the first Mariner 4 frames.) Some landscapes unique to Mars were seen in a few of the pictures from Mariners 6 and 7, flown in 1969 with much improved cameras. Still, Mariners 4, 6, and 7 together photographed only a small percentage of the Martian surface at close range, so it could not have been suspected that the few unusual regions would turn out to be the headwaters of an immense system of now-dry river valleys on Mars.

With its sister spacecraft, Mariner 8, unfortunately at the bottom of the Atlantic Ocean, Mariner 9 arrived at Mars on November 14, 1971, to find the entire planet hidden beneath a pall of dust. Bruce Murray and the Mariner 9 television research team stared gloomily at each successive blank photograph from their Mars-orbiting television camera. They had been surprised and disappointed 7 weeks earlier, when Earth-based astronomers reported what was to them a familiar happening on Mars. Telescopes revealed that a small dust cloud had grown in the Noachis area in the southern hemisphere. It had rapidly expanded and spread around the planet until in November the entire globe was shrouded. Nor was this storm mere bad luck for Mariner 9; it had been predicted to occur long before Mariner 9 had been launched.

Astronomers were bitterly familiar with the disappointments of 1956, the previous time that Mars had been simultaneously close to Earth and at the closest point to the sun in its elongated orbit. Mars can be studied every 2 years from Earth, but it is much closer on special occasions every 15 or 17 years when it is also closest to the sun. Elaborate plans to scrutinize Mars during the 1956 opportunity were thwarted by a planetwide dust storm, just as observations had been impaired during every previous such occasion back to the 1800s. Observers had long supposed that the extra solar heating of the southern hemisphere, tipped sunward at such times, generated the storms that scoured pockets of yellow dust from around the dark regions that gird the southern temperate latitudes of Mars.

If the geologists had failed to heed the astronomers' warnings in planning the strategy for Mariner 9, they had nevertheless designed a flexible mission that could outlast any dust storm. Thus the great bonus afforded atmospheric specialists, who traced Martian winds from the changing dust cloud patterns, did not come at the expense of the geologists who patiently waited for the storm to subside. They then began a photographic sequence, lasting into autumn 1972, that succeeded in mapping the whole planet.

As the dust gradually settled, Mars performed an excruciatingly prolonged striptease and gradually revealed surprises that rivaled its "moonscape" exhibition of 1965. Four huge, dark mountains poked through the dust-laden atmosphere. Each was crowned by interlocking craters that resembled not meteor-impact craters but rather the volcanic crater or caldera atop the Kilauea volcano in Hawaii. The largest of these volcanoes, Olympus Mons, is the size of the state of Colorado and exceeds in size any volcano on Earth.

The second astonishing piece of scenery to be revealed was the giant chasm later christened Valles Marineris. It is the length of the United States and so vast that one of its tributary canyons almost dwarfs the Grand Canyon. Wider than the Grand Canyon is long, Valles Marineris is dark in color and had been called the Coprates canal by astronomers decades ago. Such extraordinary canyonlands imply Martian geological forces like those in the Earth. The moonlike terrains were but a relic from Mars's ancient past, like the old cratered lake districts of Canada. The young canyons and volcanoes reestablished Mars, like the Earth, as a geologically fascinating world.

The most portentous discovery of all came in Mariner 9 pictures of the Chryse plains near the Martian equator. They showed meandering valleys with tributaries that looked exactly like river valleys on Earth. For once, extreme caution prevailed among the geologists responsible for interpreting the pictures. They had drawn premature conclusions before and were not about to do so again on the subject of Martian water, with its charged implications for life. Early attempts to explain away the meandering dry valleys as cracks, volcanic flows, etc., invoked everything but the obvious: flowing water. Eventually, however,

Mariner 9 geologists agreed that most (if not all) flow features visible in most (if not all) of the riverlike valleys could only be due to running water. Some concluded that the networks of smaller valleys, ubiquitous in the older, equatorial highlands, formed long ago when it might actually have rained on Mars. The huge "outflow" channels, of more recent age, seemed obviously due to flooding water. These features resembled the countryside of eastern Washington State, whose contours were shaped by a prehistoric flood of unprecedented scale. Apparently there had been a diluvian epoch on Mars of unimaginably catastrophic flooding. Since the environment for running water does not exist on Mars today, Mariner 9 scientists concluded that the climate had changed, perhaps irreversibly, and Martian waters were now permanently frozen in underground reservoirs, never again to cascade through Martian canyons.

It was inevitable that Viking's assault on Mars would cause our perspectives on the Red Planet to change. And so it has. Previous global generalizations ran afoul of inflexible new facts: precise analyses of surface chemistry and a plethora of newly revealed landforms just tens of feet across. The veils that obscured Mariner 9's view of Mars's polar regions have been pierced, revealing geological activity rivaling that of the two most restless planets, Earth and Io. Dramatic evidence of permafrost and ground ice in even the tropical latitudes of Mars has led some researchers to consider that glaciers—even an Ice Age—may have figured in the history of the planet. And the tide of opinion has once again begun to turn against balmy climates and flowing rivers in the Martian past.

Mars is the keystone of comparative planetology—the multidisciplinary endeavor to understand general planetary processes and to view our own planet in its solar system context. Mars lies near the asteroidal transition zone between the inner and outer planets. Among the Earth-like planets, Mars is of middling size, smaller than Earth but larger than Jupiter's moons. It does have an atmosphere, although a very thin one. Mars's cratered highlands may date from the first billion years of the planet's existence, yet other land-forming processes are actively changing its surface even today. Unlike the constantly evolving Jovian moon Io, Mars has most of its planetary history exposed for analysis. On the other hand, its internal activity has been more complex and interesting than the temporary spurts of magma or ice volcanism that occurred long ago on now-dormant bodies like the moon, Mercury, Ganymede, and Dione.

The geology and meteorology of Mars is a balance between the familiar and the exotic. On the familiar side, pictures of Martian sand dunes are indistinguishable from aerial views of the Sahara desert. There seem to be Martian landslides and graben canyons, buttes and mesas, lava flows and cinder cones, all familiar terrains to geologists who study terrestrial deserts. Specialists in the geology of Antarctica, Siberia, and Iceland recognize some peculiar Martian landforms as having terrestrial counterparts, indicating frozen ground, glacial erosion, and sub-ice volcanism. It is reasonable that a dry, cold world like Mars might mimic our own planet's driest deserts and polar wastelands and have features like those formed during our own Ice Ages. Some Martian landscapes seem uncannily Earth-like. With dust and sandstorms on the horizon and hazy water-ice clouds overhead, you could stand on Mars next to Viking and almost believe you were in a desolate corner of the Earth.

In other ways, Mars is a bizarre, hostile world. In the summer, the remnant south polar cap is made of dry ice—frozen carbon dioxide. The planet's surface is totally devoid of organic (carbon-bearing) compounds, which shocked Viking investigators much more than the absence of life. (Carbonaceous meteorites must rain down on Mars as they do on the moon; the organic compounds they contain evidently are consumed by Martian chemistry.) For such a small world, the immensity of some Martian volcanoes, canyons, and channels is staggering. Olympus Mons, the largest volcano in the solar system, has twenty times the volume of Mauna Loa, the Earth's largest. A few channels exceed 100 kilometers in *breadth*. The atmospheric pressure on Mars is so low that water, which bathes our own world, cannot even exist in liquid form. Each night, equatorial temperatures plummet from afternoon's comparatively balmy 0°F to a bitter −150°F, far colder than it ever gets on Earth, even in Antarctica. Every Martian year, the entire planet is enshrouded by a global dust storm, raised by hurricane-speed winds. No direct sunlight penetrates to the ground for weeks or months on end.

The shapes of the land and the ocean basins on our own planet are dominated by two processes that are apparently less significant on Mars: the conveyor belt plate tectonics of the Earth's crust and the rainfall-river-ponding interaction between water and land. Each process serves both to create and destroy rocks and landforms. Plate tectonics hoists great plateaus and chains of folded mountains while splitting continents apart and perpetually recycling oceanic crust. Rains erode our mountains and flood low-lying basins with lakes in which sediments sink, eventually forming layers of rock. In Chapter 12 I delve into these dominant terrestrial geological processes and put them in a planetary context. But in certain localities on our planet, lesser processes are also significant. Even minor processes may form unique ecological niches for unusual species or permit the accumulation or exposure of precious minerals, ores, and other valuable natural resources. Such secondary processes may be better understood by studying their occurrence on worlds whose geology is not overwhelmed by plate tectonics and rainfall. Mars is such a world. Practical applications will not mature until the future, but already Viking has taught us much about the Martian style of geology dominated by volcanoes, ground-ice, and wind.

Ten percent—or perhaps as much as a half—of the surface of Mars is occupied by volcanoes and vast flood-sheets of now-solidified lava. (The uncertainty is because many Martian plains lack such obvious signs of volcanism as cinder cones or overlapping flow fronts, perhaps but not necessarily due to subsequent erosion). Volcanism has shaped much of the Martian landscape from early epochs until comparatively recently. Isolated eruptions may still be occurring on Mars today. Certainly magma has surfaced on Mars in greater quantities and for longer durations than on the moon.

On Earth, volcanic landforms are much less in evidence than on Mars. They are so rare that in America many have been set aside in parks as natural wonders (e.g., the Craters of the Moon National Monument in Idaho). Mars is not actually more volcanically active than Earth. Much terrestrial volcanism occurs, invisibly, beneath the sea. Also continental volcanoes and lava plains are overwhelmed in the Earth's rock stratigraphy by rapid erosion and by the prolific formation of sedimentary rocks. While Martian volcanism may not really challenge our own planet's production of magma, surprisingly diverse volcanic landforms do exist on Mars. They range from lunarlike lava plains, pit craters, and low lava domes to bigger, more complex structures revealing long, episodic, and evolving histories. The largest, most common Martian volcanoes are extensive but comparatively gentle-sloped accumulations of basaltic flows, similar to Hawaiian and Icelandic "shield" volcanoes. (Basalts are dark rocks formed from hot magmas that originate at great depth and flow onto planetary surfaces.)

A few volcanoes on Mars look like the most common volcanoes on Earth: steep-sloped, pitted, conical mountains with rough surface texture. These volcanoes are absent on other planets such as the moon, Mercury, and Io. On Earth, most are located along continental margins and are associated with the melting of crustal plates being subducted (submerged) into the upper mantle beneath the continents. The resulting magmas are more acidic, volatile, and buoyant and are less fluid than normal basaltic lavas. Eruptions yield explosions, ash deposits, and steep-walled volcanic mountains (such as Mount St. Helens). The absence of such volcanoes on most planets and their rarity on Mars is one indication that plate tectonics may be chiefly a terrestrial phenomenon. Yet the presence of at least a few steep volcanoes on Mars, mainly dating from long ago, suggests that Martian crustal evolution may once have been more Earth-like.

Some geologists interested in Martian volcanism have been taking field trips to Iceland. It is a cold island, with large remnants of the glaciers and ice sheets that totally covered it during the last Ice Age. It is also one of the most volcanically active areas on Earth: a new volcanic island, Surtsey, formed offshore just two decades ago. Some permanently frozen ground (permafrost) exists in Iceland, as it is thought to on Mars. As on Mars, some Icelandic soils consist of windblown dust, known as loess, which was deposited around the Earth's mid-latitudes during the cold, drier climates that prevailed as Ice Age glaciers retreated. While Iceland's environment never approached Martian extremes, examination of the Icelandic interplay between volcanism and frozen terrain has provided fresh insights concerning Mars. Carlton Allen, of the University of New Mexico, believes that certain flat-topped domes in the northern mid-latitudes of Mars look just like Icelandic volcanoes known to have formed underneath once-extensive ice sheets. It may be that the vast northern plains of Mars were once covered by ice, or an ice-rich layer of dust and rock, under which volcanoes erupted, forming the Icelandic table-mountain profile. Now the underlying topography has been revealed by exhumation, as the upper layer was somehow stripped away and disappeared.

Water, one of the four classical "elements" of which the world was once believed to be made, is a remarkable substance. It remains a liquid over a far wider temperature range than do most compounds. As its phase changes from solid to liquid to vapor, its behavior is fundamentally altered. As liquid or vapor it roams freely through a planet's atmosphere and subterranean pores and aquifers (water tables). But as a solid, water-ice can be virtually immobilized, especially if it is incorporated into the ground. A miracle of the Earth's environment is that water is abundant and changes from liquid to solid at the surface or just above it in the lower atmosphere. Even in balmy Hawaii, we can ascend 13,000 feet and find snow and ice nearly year round.

A crucial difference on Mars is that the transition between ice and water's more mobile phases occurs *just beneath* the surface. If mobile phases of water ever approach the surfaces of still colder bodies, like Ganymede, they do so rarely and in small localities—like molten rock, or magma, does on Earth. But on Mars, the depth of the permafrost is just a few kilometers, which is the same scale as the surface topography. That means that faults, volcanoes, landslides, and cratering impacts can expose subsurface rocks and soils that may be saturated with liquid water. Even the permafrost itself is gradually "dried out" as water molecules sublimate from the ice into the Martian air. In fact, at times, surface temperatures on certain parts of Mars rise briefly above the freezing point of water. If Martian waters are briny salt-solutions, as seems plausible, they may occasionally be stable even at the surface of the planet and certainly might be fluid not far beneath the ground.

Analysis of Viking pictures has uncovered plenty of evidence for permafrost, water-rich terrains, surface deposits of water-ice, and, of course, the apparent rivers and streams of the past. Craters on Mars, unlike those on other rocky worlds, are surrounded by great splash deposits, with shapes ranging from flower petals to cow pats. These ubiquitous, so-called "splosh" craters, which seem to exist on icy Ganymede also, indicate that much of the Martian upper crust is saturated with water and/or water-ice. There are many eroded cliffs and buttes on Mars reminiscent of back-wasted tablelands in the American Southwest. But instead of being eroded by wind and rain following chemical weathering, Martian tablelands may consist of cemented ice-dust "rocks" that weaken as they are devolatilized, causing progressive recession of scarp faces. Perhaps aquifers occasionally are exposed or erupting magma melts the permafrost, causing Martian mudslides. Thus the proximity of Martian surface temperatures to the freezing point of its abundant water may be as important to that planet as to our own, even though the boundary temperature occurs just below ground level rather than above it. It is one more way in which Mars is intermediate between our watery world and the permanently frozen ice-moons that orbit the outer gas giants Jupiter and Saturn.

Most of the enduring Martian landforms were shaped by internal forces that heated and fractured the crust, mediated by the ice-rich surface rocks. Yet the most active geological force on Mars, which visibly modifies the surface from day to day, is the wind. There are sand drifts within a stone's throw of the Viking 1 lander (now called the Mutch Memorial Station). Viking Orbiter cameras showed that swaths of land stretching tens to hundreds of kilometers in the lee of craters and mountains were denuded and darkened as bright dust was swept away by turbulent winds during the annual planetwide dust storms. Bright dust deposits appeared elsewhere. The cameras also discovered vast dune fields surrounding Mars's north polar ice cap—a sea of sand larger than the state of Texas, equaling the most extensive dune fields on Earth, those in northern Africa.

Despite pervasive winds, the deposits of dust and sand are but transient features on the face of Mars. Wind leaves more enduring evidence by eroding craters, canyons, volcanoes, and plains made of strong rock. Is abrasion and sandblasting by windblown particles a dominant erosive process on Mars? Debate has raged on this question for years. Extensive regions on Mars seem to show rock formations characteristic of the windswept, sandblasted topography of the Earth's driest deserts. For example, Mars may have yardangs: elongated rocky hills, eroded by blowing sands into the shape of an inverted ship hull. But similar-shaped hills, called drumlins, were formed in places like upstate New York by slowly flowing glacial ice sheets. Certainly Martian winds are strong enough to initiate sandstorms which, if they were frequent and widespread, could have eroded away all Martian topography long ago. But obviously they have not done so, for many ancient craters are minimally eroded. Perhaps parts of Mars were protected from erosion over the ages by layers of ice or dust, from which they were only recently exhumed. Or maybe atmospheric conditions normally cannot generate sandstorms on Mars.

A critical question concerns the nature, origin, and quantities of fine particles on Mars. It may be that strong, abrasive sands are present in only limited quantity in certain Martian provinces and cannot migrate to erode landforms around the globe. Does sufficient dust exist on Mars to accumulate into thick sedimentary layers? Has dust been left over from the formation of Mars, or did subsequent thermal evolution convert the entire planet into rock? If the latter is true, as seems likely, then rock might have been converted back into dust by pulverizing meteorite impacts, by chemical weathering, by crystallization of salts, or by erosion of rocks by water and wind. Of course, for wind to reduce rocks to sand, there must be some sand to start with.

The observed dust clouds and sand dunes constitute but a small veneer of particles, continually migrating about the planet. But indirect evidence suggests there may be kilometers of particulates layered out in polar latitudes and bound up with water-ice around the planet. The "dust budget" of Mars—how much there is, how it was produced, how it migrates, and where it is stored—will be understood only after years of further research. There can be little doubt, however, that ice and dust together constitute a reservoir of material, most of it temporarily frozen and immobilized, which has enormous potential for modifying the appearance of the Martian landscape if it is ever liberated by climatic change or subterranean heating, as may have happened in the past. Since many researchers believe that such layers of dust and ice can store enormous amounts of gases, such as carbon dioxide, the evolution of the Martian atmosphere may also be intimately connected with that of the dust and permafrost.

So Mars is a world layered with lava, dust, and ice. Was it ever also a world with lakes and streams? The popular view, indeed the popular hope, is that the answer is "yes." But photogeologists studying pictures of Mars are having some second thoughts. If processes as different as sandstorms and Ice Ages can produce yardang-shaped ridges, how can we be sure about what has shaped the Martian surface? Certainly the "river valley" networks in the cratered highlands and the larger "outflow" flood channels have been shaped by the movement of fluid. But what was the fluid and how rapid the flow? And was the flow on the surface or underground?

Each of the enormous flood channels is said to have been scoured in perhaps a few days by a flood of water perhaps half a kilometer deep and a hundred kilometers wide. That's ten thousand times the peak water discharge of the largest recorded flood of the Mississippi! It's truly mind-boggling. Where did the water come from all at once? Where was all the eroded debris deposited? The channels commence at full width from collapsed badlands called "chaotic terrain," but the collapse depressions aren't big enough to have held the required volume of water. It has been suggested that pressure had built up in the whole regional water table, which was suddenly released, perhaps by a catastrophic cratering impact that punctured the permafrost layer on top.

Another suggestion is that the outflow channels might have been carved by fluidized rock or mud flows. Baerbel Lucchitta, of the U.S. Geological Survey, has found what she believes are telltale signs of glacial scouring of the channels. The channels might have been shaped not by catastrophic floods but by the slow movement of ice across the surface of Mars. Still other scientists have proposed that streamlines in the channels are due to winds, with the abrading sands derived preferentially from the chaotic badlands. Thus perhaps neither liquid nor solid water played any role at all in forming the largest channels on Mars!

Whatever the origin of the flood channels, the networks of smaller, interconnected valleys have more profound implications if they mean that rain once fell on Mars. David Pieri, a tall, dark-haired young geologist, began studying the valleys as a student of Carl Sagan. His early results showed the networks to be prevalent in equatorial latitudes, which suggested that warmer, rainy climates might have been restricted to the Martian tropics. But Pieri's continuing research led him inexorably to conclude that the valleys cannot be due to drainage of rainfall. Their similarities to the intricately textured, multifingered drainage basins on Earth are superficial at best. Instead they look like small, steep-sided canyons produced when resistant rock layers are undermined by seepage and underground aquifers. This process of "sapping," well known in such places as Utah, results in U-shaped or theater-shaped tributary heads, rather than in a progressively finer network of shallow rainfall drainage valleys. The idea that sapping might be an important land-shaping process on Mars was first suggested years ago by Caltech geomorphologist Robert Sharp, but it has taken detailed analysis of Viking's exquisite pictures for the idea to become widely accepted. The highland Martian valleys are all old and are no longer extending themselves by headward sapping. Perhaps that is because the aquifers have all drained, or maybe such shallow water tables have frozen now that the Martian climate has grown colder.

If David Pieri is right, it is one more proof that the subsurface transition between ice and liquid water has controlled the surface topography on Mars. Other putative river valleys on Mars are larger than Pieri's valleys, but smaller than the outflow floodways. Some seem to be deeply incised canyons also formed by sapping, and still others may have resulted from the undermining of layers of ice-dust rather than of hard rock. Still, there are a few that seem unmistakably like giant meandering rivers. Maybe they too record the traverses of ancient Martian glaciers, but more likely they once contained flowing water or brine. Perhaps these were great underground rivers—now exhumed—which once drained the aquifers that undermined the small highland canyons.

Mars appears after Viking to be a profoundly layered planet. In addition to layers of igneous rocks akin to those of Earth, layers of permafrost and ice-dust conglomerate have been draped across the Red Planet. The underlying water table has normally been trapped beneath these ice-rich sediments, which behave like rock because of the frigid temperatures near Mars's surface. Perhaps water-mediated chemistry formed a particularly impervious layer of rock beneath the water table. Whenever Marsquake fractures, impacts, and volcanic eruptions broke through the frozen layers, they were locally melted or exposed and they disintegrated. Any remaining dust component was then blown about by the wind until it was once again trapped within an icy matrix, perhaps mainly near the Martian poles where the crater-free surfaces are evidently repeatedly covered by freshly deposited plains.

That would be the uniformitarian view of how Mars might have evolved its remarkably diverse surface features without any need to postulate an especially different past climate. But inherent in this new Viking perspective of Mars is an element of instability: if even these ongoing arctic processes can form the fascinating Mars-scapes pictured by Viking, and if the Martian surface layers depend so precariously on the near-surface freezing point of water, think what just a bit of climatic warming—in the past or future—might mean for Mars!

12

The Earth in Its Planetary Context

Scientists still do not appear to understand sufficiently that all earth sciences must contribute evidence towards unveiling the state of our planet in earlier times, and that the truth of the matter can only be reached by combining all this evidence.

—*Alfred Wegener, 1928*

During the 1968 Christmas holidays, astronauts Frank Borman, James Lovell, and William Anders looked home from the vicinity of the moon and saw a cloud-bedecked, bluish planet, slowly turning in space. The Earth was seen, literally, from a new perspective. What could not have been predicted when the 1960s began was that by the time men reached the moon, the Earth would be freshly perceived and understood as a planet in an even more fundamental way, from a radically new intellectual perspective achieved through oceanographic exploration and a synthesis of many sciences. During the same decade when spacecraft were first discovering unexpected characteristics of the other planets, comfortable views about our own planet's intrinsic nature were also being turned inside out.

The emergence of continental drift and sea floor spreading—the new "global plate tectonics"—during the 1960s ranks as one of the greatest revolutions in the history of science, inasmuch as it totally undercut the cherished hypotheses of generations of geologists and geophysicists. Geology was a venerable science which had matured from a stormy youth over a century ago. By the early 1900s no science was so replete with observed fact and with adopted interpretations. For another half century the contours of the hills were measured, multitudinous fossils were classified, seismograms of earthquakes recorded, and still the fundamental assumptions of geology remained unchanged.

Fifty years after Alfred Wegener proposed that continents drift, I was taught college geology in the catacombs of the Harvard University Museum. The yellowed paintings of William Morris Davis and other past giants of Harvard's renowned Geology Department hung there to remind us—teachers and students alike—of the revered traditions of a secure science. I learned how mountains grew and how they eroded. I learned that the fixed continents grew outward from central cores by "accretion." Sediments accumulated in "geosynclines" and were somehow thrust upward to form mountains during episodes called "orogenies." The concepts were laced with jargon, which I learned, but I never quite realized that my own vague confusion about what was making it all work stemmed from the fact that nobody really knew.

The new concept of plate tectonics has brought together many diverse subjects in the study of our planet, including the growth of continents, building of mountains, deposition of mineral ores, development of sea floor ridges, seismicity, changing sea levels, and the climatology of ancient epochs. Alfred Wegener wrote in the 1920s, long before the general acceptance of his idea: "We may assume one thing as certain: The forces which displace continents are the same as those which produce great fold-mountain ranges. Continental drift, faults and compressions, earthquakes, volcanicity, marine transgression cycles [of the oceans onto land], and polar wandering are undoubtedly connected causally on a grand scale. Their common intensification in certain periods of the Earth's history shows this to be true. However, what is cause and what effect, only the future will unveil."

Global plate tectonics is the synthesis that Wegener sought, although the underlying forces within the Earth that keep the crust churning about are still only poorly understood. Imagine a lake, frozen over with ice on a frigid winter day. Then imagine a mighty rotor beneath the ice, churning the water vigorously, breaking the ice sheet into separate sections, or plates. The jostling might thrust one plate beneath another. Water would spurt up to fill gaps between plates, only to freeze and become fresh ice. The analogy is far from exact, yet Wegener's inspiration may have come from watching ice floes during his Greenland expeditions. The analogy may be even more relevant for several of the satellites of Jupiter, which have icy crusts and may have had liquid water mantles.

On the Earth the crustal plates are spherical shells, like fractured chips of an egg shell, some 70 kilometers thick (approximately 44 miles). There are half a dozen major plates and several smaller ones. Some plates are entirely of ocean floor, while others contain continents as well. A plate of ocean floor moves along like a conveyor belt, spreading from a central crack through which magmas erupt and form fresh crustal rocks. As it converges on another plate, one plate slowly dives under the other, carrying the ocean floor rocks to depths of many hundreds of kilometers into the high-pressure furnace of the Earth's mantle, where they are consumed. At such boundaries between plates are found the deepest ocean floor trenches, which have been buckled downward. Immediately landward of the trenches—over the descending and melting plate—volcanoes frequently form. It is also not surprising that most earthquakes happen near plate boundaries, where one plate is being subducted (thrust underneath another).

Continents are made of relatively light rock which tends to float on the denser rocks that underlie the oceans. Continents are too buoyant to sink into the mantle, so if converging plates bring two continents together, neither submerges and they smash into each other. The buckling produces great crumpled mountain chains, such as the Alps and the Himalayas. Over the eons new plate boundaries often form, sometimes in the middle of continents. South America and Africa were parts of a single southern hemisphere supercontinent, Gondwanaland, until it split apart about 130 million years ago to form the South Atlantic Ocean. The Red Sea and the Gulf of California are very young oceans and the valleys in East Africa may mark the beginnings of another. Meanwhile, the Pacific is narrowing. Although all these motions are measured in only centimeters per year, when the displacements accumulate for millions and billions of years they amount to many complete generations of the recycling ocean floor. The oldest oceanic crust known is less than 5 percent of the Earth's age, whereas continents have grown from ancient cores whose rocks date back more than 3½ billion years.

All that I have described is now accepted fact by geologists and geophysicists. Yet it was heresy only two decades ago, despite the abundant evidence for continental drift amassed by Alfred Wegener in 1912. Although best remembered for fitting together the jigsaw-puzzle shapes of South America and Africa, Wegener did not simply make a lucky guess, ahead of his time. The final 1929 edition of his book *Die Entstehung der Kontinente und Ozeane* (*The Origin of Continents and Oceans*) is a stupendous intellectual achievement. His evidence came from the disciplines of geodesy, geophysics, geology, paleontology, and paleoclimatology. He showed that ancient climates revealed by fossils made sense only if the continents had shifted with respect to the equator, the poles, and themselves. He showed that both rock units and fossil organisms matched between continents now separated by vast oceans. Wegener's reconstruction of Gondwanaland from Africa, South America, India, Australia, Antarctica, and Madagascar agrees *in detail* with the latest thinking. Virtually all his evidence now seems to have been generally sound, as are many of his arguments as to why the then-accepted theories were wrong. (Only his geodetic measurement of Greenland's actual ongoing drift seems to have been based on flimsy data.)

Why then did it take several more decades of stumbling down the wrong path for geologists to come around? It is not that Wegener's work was obscure and ignored at the time; his book was translated into several languages and received wide discussion and debate in the 1920s. The answer lies in the sociological nature of science. When a scientific hypothesis or model seems to work and satisfy many diverse observations, it eventually comes to be accepted as a working foundation, or paradigm. Researchers who have spent years working from such a foundation cannot casually throw it away because a single contradiction appears. It is far simpler and less unsettling to doubt the new evidence. It takes a massive amount of contrary evidence to overturn dogma in a well-developed science such as geology.

It was realized even in the 1920s that Wegener made an impressive case for continental drift. But drift was antithetical to so many cherished assumptions that when a single flaw was perceived in the theory, it was sufficient to scuttle it entirely in the minds of the majority. The flaw was mainly the inadequacy of his guesses as to what forces were driving the continental motions. He argued ineffectually that doubts about the motive force should not override the direct observations that drift was actually occurring. By the late 1950s evidence for continental drift was so overwhelming that it seems geologists adhered blindly to tradition despite persuasive evidence. Finally, in the early 1960s incontrovertible proof of spreading sea floors was found in magnetic anomalies in ocean floor rocks; and a few years later the revolution occurred as everyone jumped belatedly on the bandwagon.

The revolution in Earth science is a classic example of a process defined beforehand by Thomas Kuhn in his book *The Structure of Scientific Revolutions*. Kuhn realized the inevitability and necessity that scientists would adhere to accepted dogmas until overwhelming contrary evidence was amassed. Otherwise, if scientists were open to every alternative, their research would be disorganized and misdirected. As Kuhn wrote, "The price of significant scientific advance is a commitment that runs the risk of being wrong."

Still, if geologists had been persuaded to evaluate more objectively Wegener's arguments, we might now know much more about the Earth than we do. During the three decades from 1930 through 1950, only a handful of researchers pursued studies that had a direct bearing on what we now regard as the fundamental questions. Most scientists could not know that they were advancing geological understanding only tangentially by establishing the contradictory framework that made revolution inevitable, and indeed overdue, by the 1960s.

With this extraordinary example in mind, it is especially important that planetary scientists choose their paradigms with care. So far, with burgeoning new information about the planets, few scientists have developed commitments to any particular point of view. That is why there have been so many flip-flops in our planetary conceptions. But as the new data from moon rocks and spacecraft become digested and synthesized, there will be attempts to fashion scenarios and models that bring everything together. This process is already beginning. As researchers, we must be careful not to jump on any bandwagon before we are quite sure that we're headed down a fruitful path. For if we too hastily adopt the wrong paradigm and start the psychologically inevitable habit of closing ourselves off to plausible alternatives, decades later we may find ourselves far down the wrong road. Since science is a cultural activity, the responsibility for keeping our planetary perspectives broad will be as much that of governmental funding agencies, science teachers, and the lay public as that of planetologists themselves.

Our new perspective on the Earth's geology has had inevitable repercussions for our developing views of other rocky worlds. Some elements of Earth's geology that are rooted in plate tectonics, including giant rift canyons and volcano chains, exist on Mars as well. Yet others are absent, such as chains of folded mountains. When planetologists first examined the variety of landforms in the Mariner 9 photographs, they looked for earthly counterparts. Now that scientists are synthesizing all the Viking data and developing a global picture of Martian geology, conceptual analogs are being sought in the geology and geophysics of our own planet.

Mars is a world divided into two parts along an equator that is tilted 35 degrees with respect to the rotational equator. The cap of the planet containing most of the dark patches is covered with many craters. The other half of the planet lies at lower elevations and is mainly a vast plain, only lightly cratered. Rising from these plains is a great continent-sized dome, topped by the largest volcanoes. The more chaotic and picturesque landscapes are generally near the border between the two hemispheres, including the great Valles Marineris canyon and many of the largest channels. It is clear from studies of the mesas and buttes perched "offshore" from the margins of contiguous, cratered highlands that the latter are gradually being—or have been—undermined and eaten away by erosive processes working on the cliffs that mark the highland margins.

Inasmuch as 200 million years ago all the Earth's continents constituted a vast single continent, Pangaea (which later split into Gondwanaland and a northern part, Laurasia), it is tempting to regard the Martian cratered highlands as a continent and the low-lying plains as an ocean basin. (It is conceivable that waters once ran from the channels and ponded in some lowlands, but it is doubtful that we will ever find ocean sediments buried beneath the Martian plains.) Why does Mars (and the moon, too) have a hemispherical asymmetry? It has been speculated that before the Martian core formed, the whole interior was slowly boiling in one great convection cell, which stretched and tore asunder the original northern crust. If that was ever true of Earth, mantle convection has since evolved into a more complex pattern. But the Martian dichotomy may still shed light on the origin of our own continents.

The most intriguing region on Mars is that of the stupendous volcano-topped Tharsis dome and adjoining channel-cut lowlands to the east. The great canyon Valles Marineris divides the plateau that joins these two provinces. Some scientists are tempted to consider this region as one of incipient plate tectonics on Mars, since similar uplifts and rifts mark the beginnings of continental separation in East Africa. Perhaps Tharsis was thrust up by a mantle "plume" (the rising side of a convection cell). Associated splitting of nearby highlands may have created Valles Marineris and marked the beginnings of plate motions on Mars. Subsequently the interior of Mars cooled, the crust thickened, and the planet contracted, arresting further plate development. Certainly the heavy load of the weighty volcanoes on the Tharsis dome, which have been found—surprisingly—*not* to be simply floating in the Martian mantle, must be sustained by an extraordinarily thick, strong crust under compressional stress due to global contraction. (Cracking due to the great weight of Tharsis is an alternative explanation for canyons such as Valles Marineris.) In developing plate tectonics, Mars seems to be a world that tried to but couldn't, due to its thick crust, lack of sufficiently vigorous internal churnings, and possible lack of a sufficiently slippery boundary between its crust and mantle.

Depending on the latter point, it may be that the entire crust has slipped around on the mantle, like an egg's shell slips around the insides when it is spun up and suddenly stopped. There are hints that Mars may actually have one such spherical crustal "plate." Consider the long chain of volcanic islands stretching across the central Pacific. The continual southeastward displacement of volcanism (20 million years from Midway Island to now-active Kilauea Crater on Hawaii) seems due to the northwestward motion of the Pacific plate across an active spot in the Earth's mantle, which heats or cracks the plate, resulting in eruptions. Could the alignment of several of the giant volcanoes on the Tharsis dome have a similar explanation? It is intriguing that the apparent displacement approximates the tilt between the Martian equator and the boundary between the highlands and northern plains. Geologist Peter Schultz of the Lunar and Planetary Institute has advocated even larger polar wanderings on Mars since pre-Tharsis epochs, based on his discovery of old tropical sedimentary deposits (he calls them "paleo-poles") similar to modern deposits that are arranged in spirals about the current poles. In fact, the laws of physics require that once an excess mass like Tharsis developed somewhere on the Martian crust, the planet must have readjusted itself by polar wandering so that the excess lies near the equator (as Tharsis does today).

In Chapter 6 I discussed how internal heat drives the geology of a planet's surface. It is not surprising that Mars, which is larger than the moon but smaller than Earth, is also intermediate in its geological complexity and activity. Planetary energy is manifested on Earth by plate tectonics, while Mars epitomizes a less vigorous planet. By analogy, Venus, virtually the same size as

Earth, should have plate tectonics, too. We cannot yet be sure, but there are enticing hints that Venus may indeed have a geological style more Earth-like than that of other rocky worlds.

The Pioneer Venus orbiter carried an unheralded radar that mapped the topography of the planet between 1978 and 1981. While it could not distinguish features smaller than about 50 miles across, the radar provided the first global view of our sister planet. Most of Venus is quite level, but there is one high Earth-like continent, called Ishtar, which is bigger than Australia and stands 3 miles above the rest of the planet. Ishtar has several mountain ranges, including the mighty Montes Maxwell, which rise nearly 7 miles into the broiling Venusian sky.

One other province on Venus deserves to be called a continent: Aphrodite. It is much more extensive than Ishtar, but it is not very high nor are its boundaries so well defined as those of terrestrial continents. Still, Aphrodite, along with its neighboring lowlands and the Beta highlands, contains some remarkable sunken rift valleys and trenches. One of these, Artemis Chasma, is semicircular in shape and bears an uncanny resemblance to several abyssal trenches off Southeast Asia where plates are being subducted into the Earth's interior. Several global patterns on Venus hint at active crustal movement to some scientists, while other researchers think Venus may for some reason be frozen in an adolescent stage of crustal development. Whichever interpretation is true, all agree that the Pioneer Venus radar has only whetted our appetites for the space mission required to provide the firm answers: the Venus Orbiting Imaging Radar, now delayed to the 1990s, at the earliest, unless Congress restores its funds or President Reagan changes his mind about killing it.

While we wait to learn whether Venus behaves in accord with its Earth-like size in the hierarchy of rocky planets, Voyager has opened our eyes to other factors just as important as size in determining crustal activity. The icy moons of Jupiter and Saturn have proven that small worlds can behave like rocky worlds many times bigger, given appropriate composition. With the melting temperature of ice only a quarter that of rock, even a small body can be heated sufficiently to exude watery "magmas" and even temporarily sustain global plate tectonics, with solid ice substituting for a rocky crust. And Io has proven that even a small rocky body can outperform the Earth if it is heated enough.

As fragile, mortal creatures living on the planet Earth, we are sustained by the solid ground beneath our feet and by the benevolent climate of the air we breathe. As inquiring, sentient beings, we wonder about the origins of our environment and its future. Just in the last 20 years, we have learned that the ground is not, after all, fixed beneath our feet but, on a geological timescale, it drifts about and is sometimes recycled within our world. What about the climate? Is it constant, save for the cycle of seasons and fluctuations of weather, or is it also impermanent? Have the climates of other planets always been as hostile as we now find them to be? Do the planets hold clues about the evolution of our own environment?

There is little doubt that the Earth's climate has evolved, although how much is a matter of conjecture. We have already seen that evolution of life on our planet was affected by climatic change and, in turn, affected the climate. It may be no coincidence that the rapid evolutionary development of our own species during the past million years happened during a rare period of wild fluctuations between Ice Ages and brief temperate spells like the present. On Mars, too, there is evidence for possible climate change. If channels are due to flowing water, or even just a higher water table, climatic warming may be implied. Furthermore, peculiar layered deposits in Martian polar latitudes imply cyclical climate changes on a timescale similar to that of our own cycle of Ice Ages.

Planetary polar processes can easily dominate global climate because the poles are usually the coldest spots on a planet's surface. Atmospheric gases and oceanic fluids move freely about a planet until they encounter freezing temperatures. When that happens, they freeze and are immobilized on or within the ground or an ice cap. Were it not for the abundant oceanic reservoir of water, which could hardly all freeze out in the arctic, Earth's polar cold-traps would soon deplete atmospheric water vapor and surface waters to very low equilibrium amounts. The atmospheric pressure on many planets, including Io and Mars, is controlled by frozen deposits in the coldest regions accessible to diffusing atmospheric molecules, generally at (or just below the surfaces of) their poles.

While poles are usually cold, the polar sunshine budget is highly irregular. At the Earth's poles, the sun shines continuously for 6 months without setting! With our planet's tilt of 23½ degrees, the sun is never high in the polar sky. But if the tilt were 45 degrees, Montreal and other places at that latitude would not only have the sun up around the clock in June but would have it straight overhead at noon! It is easy to imagine that any glacial ice sheets would not last long near Montreal under such circumstances, notwithstanding the colder, darker winters Montreal would also have to endure.

Actually the Earth's tilt varies a bit, from 22 to 24½ degrees, every few 10,000 years. Moreover, the Earth's orbital ellipticity changes, as does the orientation of the orbit, so that the Earth's distance to the sun (hence solar heating) can vary more than it does now. Sometimes Earth is closest to the sun during summer in the southern hemisphere (as is true now), sometimes during the northern summer. It was originally an idea of Alfred Wegener that the sunlight received at latitudes near Montreal's would affect the growth of ice caps and hence the Earth's climate. In the past decade it has become widely accepted that variations in the volume of polar ice during the past half million years (measured by oxygen isotope ratios in fossilized plankton in deep sea sediments) are almost entirely due to the complex oscillations of the Earth's axis and orbit.

Mars is tilted about like the Earth, though its orbit is more elongated. But William Ward, a physicist at the Jet Propulsion Laboratory, has discovered that Mars's tilt varies much more than Earth's, due to some weak but repetitious gravitational effects of distant Mercury and Venus. Its orbital ellipticity also varies widely. Many scientists believe that resulting changes in the sunlight reaching various latitudes on Mars at different seasons have important, complicated effects on water, carbon dioxide, and dust, yielding the layered polar blankets of dust and sand observed by Viking.

During epochs when Mars is not tilted much, carbon dioxide

migrates out from underground reservoirs. Dust helps condense out atmospheric carbon dioxide, which snows out and piles up into dusty dry-ice polar caps tens to hundreds of meters thick. With much less air than Mars has today, dust storms cease for tens of thousands of years until the axial tilt increases again. Then the sun begins to bake the polar caps each summer, leaving dust-rich residues, observable today as the so-called layered terrains. As the poles are warmed, water-ice briefly thaws, and the Martian atmosphere is regenerated, to perhaps two or three times its current pressure.

William Ward and his colleagues have discovered that Mars's axial tilts may have been even greater in the past, before the Tharsis dome developed. Mars wobbles like a spinning top; the new Tharsis "bump" on the Martian equator made the wobble faster. But before Tharsis, the slower wobbling may have been in perfect repetition, or resonance, with the aforementioned periodicities involving the orbits of Mercury, Venus, and Mars. Occasionally Mars's tilt may have exceeded 45 degrees (as in my "let's pretend" discussion of Montreal). Probably any future pre-Tharsis-like tilt could not generate an Earth-like atmosphere on Mars. But in the pre-Tharsis past, before carbon dioxide became so thoroughly imbedded within the Martian ground, there might have been enough of an atmospheric "greenhouse" produced to warm Mars, at least briefly, above the freezing point of water so that rivers might have scoured out the channels.

Evidently changes in sunlight received by a planet, as its orientation and distance change, govern climatic and glacial cycles. Clearly, if the sun itself were to change in brightness, the climates of Earth and Mars might change in unison instead of independently. Indeed, there have been mysterious long-term climate trends on Earth that are conceivably due to changes in the sun. While the youngest Martian polar deposits are only a million years old, there is a sedimentary record on Mars extending back perhaps billions of years (for example, the "paleo-poles" of Peter Schultz). Perhaps we can learn more about the evolution of the Earth's climate from studying Mars than from our own planet's geological record, which has been so badly disrupted and distorted by the great crustal turmoil of plate tectonics.

13

The Galileo Project and the Future of Planetary Science

MR. ROYBAL. *Mr. Speaker . . . the Jupiter Orbiter Probe . . . is the product of years of engineering and scientific preparation. Many planetary missions are justifiable, but to make the best possible use of our resources, NASA, in collaboration with science and industry, selects only the best flight opportunities.*

MR. BOLAND. *Mr. Speaker . . . Jupiter will be up there 5 years from now; it will be up there 10 years from now. . . . With the launch windows that will occur again, there is no reason in the world why we have to put it in the fiscal year 1978 budget.*

—*Congressional Record (House), July 19, 1977*

I was exhilarated the morning of October 12, 1977, as I walked into the entrance of the Jet Propulsion Laboratory in Pasadena, California. Just a few weeks earlier I had received a telegram telling me my proposal to work on the Galileo Imaging Team had been accepted. Until that day, Galileo had been a paper mission as it gradually took shape in countless advisory committee reports, engineering documents, and technical proposals like my own and as its budget was sifted through the political process. Now, a new fiscal year had begun and the future was beginning to materialize. Even as I signed in and received my badge, overhead monitors displayed live network telecasts of an early test landing of the Space Shuttle, the reusable workhorse that, in a few short years, would carry Galileo into space and send it toward a 1984 rendezvous with Jupiter. I could begin to believe that my family and I would be temporarily living in Pasadena, while I studied Galileo's new pictures and helped target its cameras on the giant planet's swirling clouds as the spacecraft swung from moon to moon in a twenty-month tour of the Jupiter system.

I joined the scientists congregating in the lobby of JPL's Von Karman Auditorium. I greeted some old friends. Others were strangers wearing badges with familiar names. This first meeting of the Project would be the only full gathering of the 100-plus scientists and all the JPL managers and aerospace engineers who would develop this ambitious foray into the cosmos. For the next decade we would work primarily in small teams on different spacecraft systems and instruments.

We watched television monitors as the Space Shuttle *Enterprise* plunged in its powerless glide toward a successful desert landing 60 miles to the northeast of JPL. Other JPL monitors were flashing live Viking pictures from the surface of Mars. Outside, a large display board showed the near-Earth positions of the two Voyager spacecraft launched only weeks earlier toward the outer solar system. I had a foreboding sense that too much was happening at once, that the universe of political reality was too small to hold all these projects at the same time.

Following a brief welcome, we were introduced to John Casani. Casani had been leading the Voyager Project. But with the Voyagers safely launched, early difficulties with Galileo now required that JPL's best management talent be reassigned. So John Casani began his job as Project Manager for Galileo by telling us of a "weight problem." Early estimates were faulty, he said, and the spacecraft literally would never get off the ground. He outlined three unpleasant strategies for solving the problem: (1) reduce the mass of the Galileo orbiter, which might mean off-loading some of the scientists' recently selected experiments; (2) shorten the orbital tour and visit fewer of Jupiter's moons; or (3) "increase launch vehicle performance"—in other words, ask people in another branch of NASA to build an even better Shuttle than they had planned. The third alternative would eventually seem bitterly ironic, as the Shuttle program was plagued with development problems and delays. Four years later, instead of shipping Galileo to Florida for launch, we would still be facing four years to launch.

The delays could be tolerated. Ingenious JPL engineers could, and did, devise shortcuts to the weight problem as fast as the Shuttle's projected launch capability dwindled. But the Shuttle's cost overruns, especially in an inflationary epoch, were beyond NASA's capacity to control. The Shuttle would ultimately consume a great deal of the agency's budget, with the irrational result that the payloads would be sacrificed to save the launch vehicle. The eventuality had not been unforeseen. In the early 1970s, ex-scientist-astronaut Brian O'Leary had campaigned vigorously to postpone the Shuttle, arguing that it would starve the projects, including space science missions, that were NASA's true purpose. To ensure a constituency, during a period of public skepticism, for the Shuttle Project—so lucrative for some aerospace corporate giants—NASA decisively burned its bridges behind it. After the Voyager launches in 1977, NASA permanently dismantled its reliable Titan-Centaur launch capability. Subsequent spacecraft would be launched by the Shuttle, or not at all. For years longer than agency officials expected, it was to be not at all. Congress reluctantly subsidized the Shuttle's mounting overruns, but took its pound of flesh—and more—from the space science programs, just as the once-ignored O'Leary had predicted.

One day in February 1981, my telephone rang and I answered it. A familiar voice said, "Brace yourself for some bad news. The mark-up of President Reagan's budget is in. Six hundred and twenty-nine million cut for NASA. Two hundred million from the Office of Space Science. No VOIR. No GRO. And Galileo is cancelled again!"

My friend offered no further translation or interpretation. None was necessary. I knew GRO stood for the would-be Gamma-

Ray Observatory. VOIR ("to see" in French) was the acronym for Venus Orbiting Imaging Radar, a satellite that may one day peer through the Venusian haze and map our sister world better than the celebrated Mariner 9 mapped Mars a decade ago. VOIR was to be the salvation of NASA's Planetary Exploration Program, which had been withering during the 5 years since the last new mission was approved—Galileo. And the president would even cancel Galileo itself, the ambitious project on which $250 million had already been spent. Just as President Carter tried to do a year before, Reagan would trim a third of the space science budget to achieve a 10 percent NASA cut. And the planetary program—NASA's most visible and spectacularly successful endeavor of the past decade, despite its meager 4 percent share of NASA's budget—would be slashed by more than half!

To martial the economic, technological, and intellectual resources required to explore the moon and planets is a task that even the strictest capitalist must agree is beyond the unassisted capacity of the private sector. It was thought that the new president's advisors agreed. In fact, encouraging word leaked from Reagan's confidants had resulted in a JPL party of celebration for the Venus and comet missions the new Administration would usher into being. Now, a few days later, this news! Gone would be not just dreams of a mission to Halley's Comet, expectations for a newly authorized Venus mission, but also the Galileo mission itself, all that remained of the debilitated program.

I would have been devastated, except that I had heard these same words too often before. Galileo had been threatened with cancellation several times, twice in the past year alone. To be sure, its launch date had slipped from 1982 to 1984, then to 1985. And the Project's financial resources had been pinched back time and time again. But wiser heads in Congress had always prevailed, encouraged by the aerospace lobbyists, space activists, and concerned citizens willing to send a telegram. So engineers were kept on JPL's payroll to build Galileo's instruments and plan its encounter with Jupiter. The planetary program had survived before and could do so again—if we responded to the threat.

Scientists and space activists got their rather amateurish political apparatus into gear. Perhaps NASA officials first learned of the impending executive action from a friendly civil servant. Or maybe it was a deliberate leak of worse-than-planned cuts to intimidate NASA into lowering its budgetary expectations. In any case, within hours a *Chicago Sun-Times* reporter was asking the chairman of the Space Science Board of the National Academy of Sciences for his reaction. Before that article even appeared on newsstands in the Windy City, indeed before NASA officials even received the official word from the president's budget director, telephones—including mine—were ringing around the country. By mid-morning the next day, hundreds of telegrams began arriving at the White House, at the Budget Office, and on the desks of key senators and congressmen known to be both sympathetic to NASA and probably influential with the new Republican administration. Aerospace executives rushed to Washington to buttonhole any appropriate official. As it turned out, the threatened cut was an exaggeration—a trial balloon. But as someone said, "When they run up the flag, you've got to shoot at it." Otherwise, by the time Congress eventually acted, the potential deletions could become bleak reality.

The planetary program is unusually vulnerable to the fickleness of the political process. Politicians often try to weather transient crises by stealing from the future. Returns from investment in the space program are indirect or intangible and long-term. Development necessary to mount a mission also takes years. Even after launch, it may be a decade before a probe reaches its planetary destination and returns data to Earth. It has seemed incongruous to the lay public and Congress that the planetary program could be in desperate straits just when newspapers were filled with marvelous new pictures of Io's erupting volcanoes and Saturn's rings. Yet the spectacular achievements of Voyager from 1979 to 1981 were initiated by Congress way back in 1971 during an era when a new planetary mission was being approved each year. Galileo, approved after a political brouhaha in the summer of 1977, will not deliver its probe into Jupiter's equatorial maelstrom until approximately 1990. The failure of Congress to authorize any other planetary missions since 1975 has *guaranteed* that there will be no planetary encounters between the second

Voyager Saturn encounter in 1981 and 1986, when the same spacecraft reaches Uranus. And congressional action is determining today whether the cupboard will continue to be bare through the 1990s.

How had we reached this juncture in exploring the solar system that at the peak of Voyager's success, the president could contemplate canceling it all? Galileo is the 1980s centerpiece of an enterprise that was conceived as science fiction, but which began to be realized in 1969 when a human being actually stepped onto another world. It is the culmination of an advisory and decision-making process that commenced nearly two decades ago. In the early 1960s, American reconnaissance of Mars and Venus was just underway, and space engineers began to think about the more distant planets, stimulated by developing scientific ideas about their importance for understanding the origin and chemistry of the solar system. The inner planets were warm and small, and hence had been depleted of hydrogen-rich gases. While that made them better abodes for life (rocky surfaces, salubrious climates), the giant outer planets had collected a better sample of the cosmic material of which stars are made and thus held unique clues to our primordial origins. Moreover, they were fascinating targets in their own right with their rings, Red Spot, and satellite systems.

Analysts working for several aerospace think tanks learned how planetary gravity fields could accelerate a spacecraft from one planet to the next. They identified how the unusual outer planet alignment that would occur in the early 1980s might help our feeble rockets take a Grand Tour of the outer reaches of the solar system. The first outer planets mission chosen was to be a simple scouting mission. Pioneers 10 and 11 would probe the asteroidal zone and Jupiter's radiation belts to test the environmental hazards later spacecraft might encounter. Congress authorized the mission in 1968, Pioneer 10 was launched in 1972, and it flew past Jupiter in late 1973.

The meaty scientific exploration mission would be the Grand Tour, to be followed by intensive study of each outer planet.

Already by the late 1960s, a Jupiter orbiter was conceived that would evolve into the Galileo mission. But the fiscal strain of a guns-and-butter economic policy and public boredom with moonwalks coalesced to put the Space Program into reverse gear. Planned moon landings were abruptly canceled and NASA was forced into its fallback version of the Grand Tour: a Jupiter-Saturn flyby using the tried-and-true Mariner spacecraft, a mission that would be called Voyager. (Voyager may be lucky and make it to Uranus and beyond, but it was not engineered to last beyond Saturn.)

At the same time, the scientific advisory system that played a role in the shaping of the Planetary Exploration Program was also evolving, becoming more structured and influential but also more politically complex and controversial. In the early 1960s, individual scientists consulted with the engineers who developed mission concepts. Then NASA created a Lunar and Planetary Missions Board and a succession of scientific "working groups" to evaluate its engineers' mission plans. NASA now receives parallel advice from its own committees and from the National Academy of Sciences, the self-governing society of leading scientists chartered by Congress to advise the federal government. Academy recommendations are developed by Complex, the *COM*mittee on *P*lanetary and *L*unar *EX*ploration of the Academy's Space Science Board. Although its members often are NASA-funded colleagues of NASA's internal science advisors, Complex developed a style of insistent independence, which heightened tension—but perhaps also the quality of advice—during the increasingly frustrating decade of the 1970s.

Scientific advice might seem central to a program whose ostensible purpose is the scientific exploration of the solar system. Yet the advisory process has been buffeted by extraneous forces as well as internal dissension. None of the advice of Complex has been put into effect during the past decade, with the single, shaky exception of the Galileo mission. Interdisciplinary jealousies have been one problem. The space science program was dominated from its earliest years by space physicists interested in measuring magnetic fields, charged particles (electrons, protons, etc.), and related phenomena in near-planetary and interplanetary space. For a while, specialists in planetary atmospheres and surfaces found themselves outpoliticked in advisory councils. Later, NASA officials recognized the public relations value of picture-taking missions, so photogeology came to the fore while *in situ* measurements struggled for recognition. As the budgetary pie grew smaller, interdisciplinary conflicts became heated. In order to gain broad backing for a mission from the whole scientific community, NASA was forced to consider larger, more complex spacecraft that would carry instruments to satisfy every discipline. It would be a scientific bonanza to synthesize so many different data, but the price tags for such ambitious missions exceeded budgetary limits and they were rarely approved.

Another dialectic has involved the balance between sheer exploration and the rational pursuit of well-posed questions. The usual *modus operandi* of science is the latter: to try to answer questions posed by knowledge in hand. Successive Mariner and Viking missions provide impetus for a new Mars mission designed to address recently recognized problems. Meanwhile, even the *first* reconnaissance missions to a comet, asteroid, or the outermost planets are postponed. Those little-studied bodies have a smaller scientific constituency. Moreover, it cannot be *assured* that important issues will be resolved by such a mission, and unknown environmental hazards risk failure. Complex has generally preferred returning to Mars and Jupiter for intensive study, despite its own eloquent expression of the minority view in its 1974 report:

> Too often we as scientists may forget that the exploration, the search of the unknown, for mystery is that precious substance upon which all science ultimately feeds. Those who cannot realize the basic value in exploring for new questions and offer only solutions for obvious questions fall short of the basic scientific goal. Clearly a balance is necessary. We must conduct sober scientific study of questions posed. . . . [But] to claim that exploration is not the motivation for the investigation of the solar system is to assert emphatically that the unknown scientific question is not worth knowing.

Scientists often fret that their advice is heeded too little and too late in NASA's development of planetary missions. In the 1960s, NASA engineers examined available spacecraft and instruments, launch vehicles and opportunities, and budgetary projections—only then did they present their array of mission options to science groups for evaluation of merit. When Gerald Wasserburg assumed the chairmanship of Complex in 1974, he tried to focus engineers' talents more sharply on scientifically rewarding missions by establishing required scientific objectives first. In its 1975 report, the Space Science Board broke with tradition and adopted the Wasserburg approach as its formal policy. No longer would scientists pick and choose among missions, with all their attendant engineering and political complexities.

Gerald Wasserburg had developed a laboratory at the California Institute of Technology for analyzing minute amounts of isotopes in moonrocks, meteorites, and terrestrial rocks in order to determine their ages, when they formed. His fastidiousness about the cleanliness of his lab led to measurements regarded as the most reliable in the business. During scientific meetings, Wasserburg's incisive mind quickly detects sloppiness in his colleagues' scientific reasoning, which he is adept at exposing with devastating directness from his seat in the back of the room. Wasserburg tries to be more diplomatic in matters of scientific politics, which results in oblique but emphatic oracular discourses, reinforced by penetrating eye-contact, that leave his first-hearers nodding in agreement but perplexed. Through repeated colloquies of this sort with groups and individuals, Wasserburg nurtured the belief that the National Academy might publicly oppose a planetary mission that failed to make the measurements required to meet Complex's science objectives. Thus he prodded NASA to consider scientific objectives more seriously.

To ensure the threat could not be undercut, Gerald Wasserburg worked painstakingly on every word of the two Complex reports issued during his chairmanship. Those exploration strategies for the inner and outer planets represent the most careful, rigorous advice ever developed by a space advisory group. To demonstrate his serious intent, Wasserburg took the further, controversial step of writing some clear (though polite) statements that could only be construed as *opposing* several popular projected planetary missions. Thus while one internal NASA advisory committee was unrealistically recommending ten missions for launch in the 1980s (and they didn't even consider missions to the outer planets), Complex's decadal strategy for the whole solar system could be met by perhaps five missions. Today, even that seems a luxury, but the committee's restraint gave its positive advice credibility and political clout. NASA paid attention, and amalgamated a Jupiter Orbiter Probe (JOP) project from parts of separate Mariner-Orbiter and Pioneer-with-Probe concepts that had been developed earlier at JPL and at NASA's Ames Research Center. The probe would actually dive through Jupiter's clouds and meet Complex's prime objectives, while the orbiter would not only relay the probe's data to Earth but would circle Jupiter for two years, monitoring the planet's weather, magnetosphere, and satellites.

In late 1976 NASA invited scientists from around the world to propose instruments to fly on JOP, later named Galileo. The advertised scientific objectives of the mission would be determination of:

> (1) the chemical composition and physical state of [Jupiter's] atmosphere; (2) the chemical composition and physical state of the satellites; and (3) the topology and behavior of the magnetic field and the energetic particle fluxes.

The objectives were verbatim from Complex, but whereas Complex had explicitly ranked the objectives "in order of importance," NASA tried a more evenhanded approach, even inverting objectives (2) and (3). NASA was influenced by (besides politics) "financial and instrumental realities." Available techniques for determining Galilean satellite compositions were borderline: some might yield ambiguous results, others might not even work near Jupiter's radiation belts. Complex had urged an expeditious research and development program to improve the techniques, but NASA wished to sidestep the issue by demoting objective (2). During ensuing years, Complex would hound NASA about the importance of the satellite objective, among others, and would demand testimony from scientists designing the satellite instruments. Some onlookers felt that Complex demeaned fellow scientists and was micromanaging the project, beyond the scope of its advisory charter. The committee felt, however, that its long-term strategies would be meaningless if fundamental objectives could be set aside in the face of routine engineering or fiscal problems. In the end, the fantastic satellite pictures from the 1979 Voyager flybys did more to whet appetites for Galileo's compositional measurements than all of Complex's exhortations.

By the summer of 1977, NASA had completed engineering blueprints for the JOP mission and had evaluated proposals from nearly 500 scientists. JOP was in President Carter's Fiscal Year 1978 budget, but at the last minute, Massachusetts Representative Edward Boland threw a monkey wrench into the process. He and his aides had become irritated with NASA's approach to priorities and launch vehicles. As the powerful chairman of a House Appropriations Subcommittee, he was able to delete funds for JOP from the House version of the NASA appropriations bill, despite Senate approval. Such conflicts between House and Senate are normally ironed out in a compromise "conference." But Boland stubbornly resisted the Senate's unyielding support for JOP, so the stage was set for an extraordinary event.

On July 19, 1977, the merits of the Planetary Program, and of JOP in particular, were debated on the floor of the full House of Representatives. Congressmen had been well versed about JOP, thanks to the indefatigable Gerald Wasserburg and hundreds of other scientists, aerospace lobbyists, and space activitists such as those in the L-5 Society (which promotes a future space colony between the Earth and the moon). Member after member rose to defend a mission that would help us, as one put it, "understand more about the universe in order that we may, in fact, know more about ourselves and our own destiny as beings on this planet." Rarely have politicians been so eloquent in defending basic scientific research. Congressmen from California, personally lobbied by Governor Jerry Brown, pinned the survival of Pasadena's Jet Propulsion Laboratory on approval of JOP. "JOP is recommended by the National Academy of Science," said one representative.

Congressman Boland argued lamely that "NASA is not going to be short-changed in planetary probes," then ticked off names of half-a-dozen projected (but unfunded!) planetary missions. He ignored the unprecedented lull in planetary exploration already mandated by the government's previous failure to approve new missions. Boland's speech was just for the record, however, for he could see how the vote would turn out, and why: "Mr. Speaker, obviously the horses are out of the barn and have been all lined up. We have California, where the Jet Propulsion Laboratory will be working on the Jupiter Orbiter Probe. We have, as practically everybody knows, the colleges and universities. . . . We also have all the contractors all over the Nation who have any part or parcel of the space program calling Members saying, 'You ought to put the $20.7 million back in for this program.' . . . The phones have been ringing off the hook." The roll was called and JOP won by 280 to 131.

NASA officials had not counted on such a large margin of victory, however, and had used some tactics that would come back to haunt them. First, in order to squeeze under zero-based budgeting guidelines, they had committed to a total cost for JOP that was tight, to say the least. Second, they had sold the dubious proposition that JOP required immediate approval to meet the unique 1982 launch window. As one congressman restated it during the floor debate, "If we do not get the program started this year, we cannot get the probe off in 1981 or 1982. And, if we do not get it off in 1981 or 1982, we will not be able to get it off at all until 1987. Then, [Jupiter] will be so far away that we will not be able to carry any meaningful instruments along. So, it is this year or not at all."

Within months NASA discovered the "weight problem" and also a serious mistake in the preliminary cost estimates, on which the commitment to Congress was based. Moreover, JOP was already planned to be run in an unusually frugal fashion: there would be only one spacecraft, with no backup. Other redundancies to protect against failure would be minimized. JOP would be a "success-oriented mission," in NASA's Orwellian jargon. There was precious little fat to be cut to save on either cost or weight.

Worse, the Space Shuttle was also a "success-oriented" project.

Its heat-resistant tiles came unglued. There was unexpected metal fatigue in the main engines. The upper stage, to be used for planetary launches, ran way over budget at Boeing. So Galileo's 1981–1982 launch date was slipped to 1984 and Representative Boland, remembering "now or never," was not amused. Complex and the NASA Planetary Office accepted assurances by Shuttle project officials that everything would work out with the Shuttle and the upper stage. But the well-briefed Boland thought otherwise; in 1979 he tried to order NASA to cancel the upper stage and substitute a more capable Centaur booster. NASA's lawyers concluded that Boland lacked authority to require the change, so they refused. A year and a half (and tens of millions of dollars) later, red-faced NASA officials belatedly acknowledged Boland's foresight and made the Centaur substitution. Further Shuttle delays, compounded by galloping inflation, boosted Galileo's cost beyond double the promised ceiling.

Buffeted by forces beyond their control, scientists looked on with helpless frustration as the planetary program ground to a halt. Complex narrowed its efforts to reminding NASA about the necessity for Galileo to achieve even its *prime* objective. Complex worried that Project managers, in cutting cost and weight, might sacrifice the entry probe's capacity to penetrate below Jupiter's water clouds and make precise measurements vital to the prime objective, which after all was a major motivation for Galileo's very existence. While the Shuttle escalated Galileo's costs by hundreds of millions of dollars, some scientists feared that such insistence on scientific integrity of the mission—which might cost a few million dollars—could be the last straw that would kill Galileo for good. Thus sparring with a political, bureaucratic, and technological nightmare, the scientific advisory process carried on.

Meanwhile, political realities in Washington grew even grimmer. Galileo temporarily survived the severe NASA budget cuts of spring 1981. But in autumn, a new round of budget cuts threatened to cancel not only Galileo, but the entire planetary program! It almost happened without debate and without a White House policy decision. President Reagan hinged the success of his economic program on the psychology of "no exceptions." Instead of looking into each agency's programs for waste and inefficiency to excise, Reagan simply applied across-the-board "belt-tightening." Apparently NASA's less-than-1-percent of the federal budget didn't warrant the separate examination due an agency that, unlike most others, had already been belt-tightening with a declining budget for a decade. But NASA officials balked and demanded that the White House make policy decisions about which of its non-Shuttle endeavors would be dismantled: space science, applications, or aeronautics.

In early 1982, the president slashed most planetary programs, yet Galileo "survived what we believe to be the final threat," as John Casani wrote to his team. Hard-ball politics was said to have been responsible. But the Centaur upper stage for the Shuttle was once more cut from the budget, and nobody knew how Congressman Boland might react. Without Centaur, Galileo could not reach Jupiter until 1990.

Thinking about future planetary endeavors, scientists worried about the attitude of the second-highest-ranking Reagan appointee in NASA, who was now working from a position of power *within* the agency. Once, as director of NASA's Ames Research Center, Hans Mark had dismantled Ames's first-rate research group in planetary geology. Now Mark was arguing that two decades of planetary exploration had yielded few intellectual fruits (he personally preferred astrophysics to earth sciences) and that planetary programs should be further de-emphasized (as if that were possible) until NASA could build a manned Space Station from which to launch a new generation of spacecraft. Scientists feared that, like the Shuttle, the Space Station project would consume all the funds, and once again a quasi-military project would freeze out—this time perhaps completely—the scientific and exploratory purposes for which the civilian agency was first created in 1958.

There had been thirty-two planetary launches in the 1960s and eleven in the 1970s. Now in 1982, scientists were reduced to desperately trying to save just two launches planned for the 1980s (Galileo and VOIR). New committees were formed and studies commissioned. But increasingly it seemed that the future of humanity's first exploratory advances into the cosmos would rest with the preferences of a few people like Hans Mark, blunt approaches to national economic woes, and sheer politics. (And with the Russians, Europeans, and Japanese.)

A golden age of planetary exploration has ended—temporarily, let's hope. Paradoxically, its termination could spark a revolution in planetary science. Researchers finally may have time to sift through all the accumulated information—the thousands of pictures filed away and the magnetic reels of unprocessed data. Alas, scientists are no longer distracted by exciting new encounters nor by much need to be planning forthcoming ones. But as instant discoveries abate in the 1980s, we may begin to gain understanding and perspective. A new comparative planetological synthesis can be developed and we can place our own planet in its cosmic context. The real value of spacecraft discoveries of sulfurous volcanoes and intertwined rings lies not in the "gee-whiz" findings alone but in finally appreciating, through laborious analysis, *why* Io has such volcanoes (and Europa doesn't) and learning *what* underlying principles govern the remarkable behavior of Saturn's rings.

Whether such an age of planetary synthesis arrives is not foreordained, however, just because we have all the data and free time. Scientific creativity requires nourishment that few people appreciate. The best scientists thrive when they are immersed in their research, evaluating and criticizing each other's work, and constantly stimulated by a steady stream of new facts and ideas. Over the centuries, science has institutionalized the creative interactions that augment insight: frequent colloquiums and meetings in which results and hypotheses are discussed and criticized, journals and magazines that publish research papers after critical peer-review and revision, and an academic system that involves students directly in research. Most scientists require a fresh infusion of new data and direct involvement in experiments or observation: an astronomer needs to observe, a geologist needs to go "into the field," a physicist or chemist needs to design instruments and make measurements. So now that new data from spacecraft have been cut off, it is essential that the other avenues of intellectual stimulation be nurtured so that the vitality of planetary science is maintained.

Of course, it all takes money. And money won't be available for long, nor will the best scientists be motivated to stay in the field, unless planetary exploration has a future—unless spacecraft, unmanned and manned, once more venture out into the solar system to explore its mysteries. In simpler, bygone centuries, private benefactors could underwrite the few researchers. Today almost all planetary scientists are employed, directly or indirectly, by taxpayers, through NASA. Even professors on state or endowed salaries require NASA funds for their computers, research assistants, laboratory equipment, and travel. And NASA support of such planetary research is predicated on an active flight program.

Even as fewer scientists are supported by flight project funds, there are pressures within NASA to decrease rather than increase the funding for the basic research and data analysis programs. As budgets dwindle, researchers must cut back on scientific meetings, reduce library subscriptions to journals, and postpone instrument development programs. With fewer research assistantships available, bright young students who could bring fresh perspectives to planetary problems are discouraged from entering the field. Government cuts in "wasteful travel" can especially cripple scientific productivity. No funds were authorized for U.S. Geological Survey geologists to travel to Mount St. Helens when it erupted! Some went anyway. But taxpayers can hardly expect scientists on meager academic salaries to pay their own way to the field trips and meetings that are the lifeblood of their profession. It will not be possible to maintain the teamwork, brilliance, and imagination necessary to continue the remarkable enterprise of planetary research without a reversal of the funding trends.

Will the unparalleled opportunity to foster productive planetary research be seized during the 1980s? Or will scientists' creative energies be drained writing unadopted proposals, laying off assistants, turning away inquisitive students, and forsaking creative interchanges with their colleagues? As I said before, the root of all these problems is funding austerity, compounded by doubt about whether planetary exploration will even have a future. Whether in the 1980s we will reap the return on our investment of a dozen years ago, when we launched an armada of planetary spacecraft, depends on the future success of the science advisory system, on the wisdom of NASA management, and ultimately on the administration and the Congress, both of which must decide whether to maintain an ample research effort during the next few years and to embark on a renewed program of space exploration.

If we can afford to rekindle the space program, fly future missions to our neighboring worlds, and pay scientists to analyze the data, we may soon gain deeper insight into our planet's history and place in the solar system. What we know about the planets, we have learned through the individual and collective perspectives of scientists. Although I am a scientist myself, the way that men and women acquire scientific wisdom remains as marvelously mysterious to me as do other complex human activities. But I know that the sociological and intellectual process called "science" is a fragile endeavor. In this volume I have tried to provide a few anecdotes and philosophical musings to help readers appreciate the subjective aspects of planetary science as well as its tentative results. I hope that it has become clear that planetary research is not a mechanical thing that just happens. Rather it requires steady nourishment and public support if it is to contribute its fruits to our culture and civilization.

It seems to me an intellectual feat of the most awe-inspiring sort that in the few centuries since Galileo invented the telescope we learned so much about the planets merely by studying their faint, shimmering light reaching us across the vastness of space. And it is a technological marvel of the Space Age that in the last decade we obtained pieces of the moon to study in our laboratories and directed remote-controlled robots millions of miles away to hammer at Martian rocks. In the early 1980s, we stand at the threshold of a broader opportunity to reach out to grasp the knowledge waiting for us on the planets and build on what we have learned so far. I wonder why we are hesitating.

Index

adsorption, 103
Alfvén, Hannes, 46
Allen, Carlton, 178
Alvarez, Luis, xiv
Alvarez, Walter, xiv–xvi
American Astronomical Society (Division for Planetary Science), 7, 94
Ames Research Center, 18, 205, 209
Antarctica, 9, 167, 175–76
Antoniadi, E. M., 67
Apollo program, 22, 51, 111–7, 123, 167, 202; reasons for, 113–5; specific missions (11–17), viii–ix, 21–22, 85, 121
Arecibo radar, 69–70, 72
asteroids and asteroid belt, xiv–xv, 8, 16, 18, 22, 24, 32, 35, 37–39, 43–62, 87, 104–105, 126–28, 141, 157, 174, 201, 203; as binaries, 43, 50; as planetesimals, 59–60; collisions among, 48–51, 55–59; composition, 51–54, 56; cores, 55–56, 60; C-types, 53–55; discovery and naming, 44–45; exploded planet hypothesis, 48; families, 49; fragmentation, 48–51, 54–57, 60; mining potential, 2, 54; missions, 32, 46; orbits and velocities, 48, 58–59; origin of, 50, 57–59, 127; sizes, 43, 47, 50, 54–56; spins and shapes, 45, 50; S-types, 54–57; thermal evolution, 55–57, 60–62; Trojan, 46; *see also* meteorites; *individual asteroid names*
astronauts, 1, 7, 21, 31, 60, 112, 115, 117, 184, 198, 201
astronomers, 2–9, 12, 40, 44, 51, 65, 71–72, 90–91, 93, 95, 157–58, 167–70, 210
atmospheres, 89–110; electricity in, 90, 96, 138, 160; escape of, 102–3, 105–6, 108–9, 154; motions and winds of, 30, 33–34, 80, 90, 97–101, 106, 173, 176; origin and development of, 102–110; *see also individual planets*
aurorae, 24, 76, 138

Bamberga (asteroid), 43, 52–53
basaltic lava. *See* volcanism
Basaltic Volcanism on the Terrestrial Planets, 123
Belton, Michael, 95, 100
Binder, Alan, 130–31, 137
biologists, 3, 166–68
Bobrovnikoff, N. T., 43
Boland, Congressman Edward, 196, 206–209
Brown, Robert A., 129, 131, 133

California Institute of Technology (Caltech), 36, 72, 169–70, 204
Callisto (Jupiter satellite), 138–43, *ills. 19–21*; composition, 138, 140; formation of, 139–40; icy crust, 139–40; impact cratering, 38, 140–43, 147; interior, 139–40; topography and thermal evolution, 140
carbon, 53, 104, 107, 175
carbonate rocks, 19, 102, 104, 107
carbon dioxide, 90–91, 93, 95–96, 102, 106–8, 175, 181, 194–95
Carter, President Jimmy, 199, 206
Casani, John, 197, 209
Cassen, Patrick, 133–36, 145–46
catastrophism, xiv, xvi, 28–29, 32, 35, 40–42
celestial mechanics, 3, 16, 36, 39, 41, 48, 58–59, 66, 71, 125, 134, 142, 159–62, 194–95, 201
Ceres (asteroid), 44, 47, 58
chemical equilibrium, 105–107
chemical reactions, 34, 59, 89–90, 95, 103, 105, 109, 126, 166, 175
chemists and geochemists, 11–12, 116, 118–19, 167, 210
chronology, geologic. *See* geochronology; impact cratering; *individual planets*
Clayton, Robert, 127
climate change. *See individual planets*
clouds, 94, 97, 100–101; *see also individual planets*
comets, 4, 18, 22, 36, 38–39, 46, 60, 91, 104, 114, 126, 141, 156–57, 199, 203
comparative planetology, 17–18, 76, 87–88, 92, 98, 101–102, 105, 123–24, 147, 174–76, 179, 191–92, 195, 210
Complex (advisory committee). *See* National Academy of Sciences
computers, 11, 81, 92, 98, 150, 162, 165, 211
conduction, 79–80, 92, 121, 139
continental drift, 15–16, 30, 32–33, 42, 78–80, 140, 147, 185, 187–88, 190–91; *see also* plate tectonics
convection, 33, 78, 80, 90, 92–93, 99, 101, 190–91
Copernicus, Nicolaus, 63
craters, 13–26, *ills. 2, 8, 9, 41*; erosion of, 23–25; *see also* impact cratering; Meteor Crater; *individual planets*
creation "science," xiv, 29
Cretaceous age, xiv–xv
Cruikshank, Dale, 130–31, 137

Darwin, George, 66, 71, 133
differentiation, geochemical, 53, 55, 60, 80–81, 83, 85–86, 117–20, 123, 126, 135, 139
dinosaurs, xiv–xvi, 21, 32
Dione (Saturn satellite), 156, 175
Dollfus, Audouin, 64, 68, 72, 157
Donahue, Thomas, 108
doppler shift, 69–70, 158

Earth, *ills. 1, 2, 11, 42*; atmosphere, xv, 9, 89–93, 97–110; atmospheric evolution, 106–7; atmospheric origin, 81, 102–5; axial tilt, 194; chronologic history, 19–20, 187; climate and climate change, xv, 27–28, 32, 34–35, 88, 97, 108–9, 178, 185, 187, 193–95; crust, 53, 124–25, 186; erosion, 15–17, 23, 29, 31, 34, 79, 90, 176–77, 182; evolution of, 10, 17, 32, 79, 81; formation of, 36, 57, 104–5, 126–27; gravity, 77, 89, 125–26;

human impact on environment, 27–28, 34, 76–77, 97, 109–110; impact cratering, xv, 13–17, 22, 32, 35–36; interior, 33, 76–79, 81–83, 90, 102, 107, 124–25, 186, 190–91; magnetic field, 76, 78, 84, 87–88; moon origin from, 124–25; new geophysical perspectives, 184–89; ocean, 81, 89–90, 98, 107, 133, 171, 185–86, 193; permafrost and loess, 178; stratosphere, 77, 100, 107; temperature, 90, 93; thermal evolution, 33, 60, 79–81; viewed from space, 184; volcanism, 9, 15, 18, 30–31, 77–78, 80–81, 90, 107, 123, 135–36, 145–46, 173, 177–78, 186, 191; weather, 33–34, 97–98; *see also* catastrophism; continental drift; erosive processes; floods; ice ages; life; ozone layer; uniformitarianism; weathering (of rocks)
earthquakes, 14, 31, 77–78, 183, 186
eclipse, lunar, 8
Einstein, Albert, 32, 41
electromagnetic radiation: infrared, 5, 52, 79, 92–93, 95–96, 108, 130, 132, 138–39; radio, 5, 68–70, 91, 130, 138, 153, 159; ultraviolet, 5, 106–107, 138, 154; visible, 51–52, 79, 92–93, 95, 102
Elsa (asteroid), 45
Enceladus (Saturn satellite), xii, 143, 156, *ill. 29*
engineers and engineering, 150, 164, 196–99, 201–202, 204, 206
erosive processes, 17, 23, 34, 176, 180; *see also* craters; *individual planets*
Europa (Jupiter satellite), 134, 143–46, *ills. 15, 16*; composition, 138, 143, 145; formation of, 139; heating and thermal evolution, 145–46; interior, 145–46; stripes, 144–46; topography, 144–47; Voyager studies, 143–44, 146; water "volcanism" model, 145–46
European space program, 168, 210
explorers, 31, 167–68

Fanale, Fraser, 137
Feierberg, Michael, 56
Finnerty, Tony, 145
floods, 27–29, 31–32, 174, 181
fossils, xiv, 19, 29, 79, 185, 187, 194
Franklin, Fred, 161–62

Gaffey, Michael, 54
Galilei, Galileo, 66, 130, 143, 147, 212
Galileo (mission), 143, 146–47, 196–202, 205–9
Ganymede (Jupiter satellite), 138–43, 145, 175, 178–79, *ills. 17, 18*; composition, 138; formation of, 139; grooved terrain and tectonics, 141–43, 147; icy crust, 139, 141–43; impact cratering, 38, 134, 140, 142–43; interior, 139–40; thermal evolution, 139, 142
Gehrels, Tom, 44–47, 56
geochronology, 19–20, 22, 36, 204
geologists, 2–7, 9–10, 29–30, 144, 167–70, 173, 175, 185, 187–88, 210
geology, history of, xiv, 19, 28–30, 185, 187–88
Grand Canyon, 29, 31, 51, 173
Great Cataclysm, 35–40, 86
Greenberg, Richard, 56
greenhouse effect, 92–96, 99, 108, 195

Hadley, George, 99
Hadley cell, 99–101
Halley's Comet, 114, 199
Hartmann, William, 21–22, 116
Hawaii: geology of, 78, 173, 175, 177–78, 191; observatories, 52, 136
heat: sources of, 32–33, 60–61, 79, 86–87, 134, 139, 145; transport of, *see* conduction; convection; electromagnetic radiation; planets, thermal evolution of; *individual planets*
Hoffman, John, 108
Hoyle, Fred, 91
Hutton, James, 29
Huygens, Christiaan, 148–49
hydrogen, 53, 102–3, 106–9, 154, 201

Iapetus (Saturn satellite), 66, *ill. 28*
Icarus (asteroid), 45
ice ages, 27, 29, 32, 34, 77, 175, 177–78, 181, 193–94
ice and ices, 34, 57, 80, 103–4, 139–40, 142–43, 159, 178–79, 192–93
Iceland, 175, 177–78
imaging from spacecraft, 6, 72–74, 150, 153, 164, 169–72, 197, 203, *ill. 3*
impact cratering, xvi, 13–26, 30, 49, *ills. 2, 19*; chronology from, 13, 19, 20–23, 25, 36, 38–39, 87, 122, 142, 144, 170; crater forming process, xv, 13–15, 18–19; late heavy bombardment, 21–22, 35–40, 86; of inner planets, 15, 17, 22, 35–39, 104, 141; of outer planet satellites, 15, 38, 140–42, 155–57; *see also* craters; *individual planets*
infrared radiation. *See* electromagnetic radiation
International Geophysical Year, 10, 82
Io (Jupiter satellite), 17, 129–138, 146, 174–75, 177, 193, 200, *ills. 13, 14*; and Jupiter's radio noise, 130, 137; atmosphere, 131, 137, 193; color, 129, 131, 136, 138; composition, 132, 138; flux tube, 137–38; formation of, 139; geology, 136; heat flow and hot spots, 132, 135–36; interior, 135, 137; plasma torus, 131–32, 137–38; post-eclipse brightening, 131, 137; pre-Voyager observations, 129–33; sodium glow, 129, 131, 133; spin, 133–34; sulfur volcanism, 130, 132, 135–38, 210; tidal heating, 80, 133–36; Voyager 1 encounter, 129, 135–36
isotopes, 20, 61, 79–80, 87, 104–5, 108, 117, 134, 204; of oxygen, 105, 126–27, 194

Japanese space program, xi, 210
Jet Propulsion Laboratory, 35, 74–75, 150–51, 165, 170, 196–99, 205–207
Johnson Space Center, 112
Johnson, Torrence, 130
Jones, Ken, 25
journalists and the media, xii, 71, 112, 150–51, 165–66, 200
Jupiter, 43, 197; atmosphere, 2, 45, 205, 208; aurorae, 138; effect on asteroid formation, 58–59; formation of, 58, 139, 201; gravity, 58, 99, 134; interactions with Io, 130–32, 137–38; internal heat, 98, 138; magnetic field, 76, 130–32, 137, 205; radio noise,

130, 137; ring, 148; satellites of, 39, 80, 124, 129–47, 157, 179, 186, 192, 205; spin, 98, 137; *see also names of satellites*
Jupiter Orbiter Probe. *See* Galileo (mission)

Kelvin, Lord, 33
Kepler, Johannes, 66, 159
kimberlite pipes, 78, 145–46
Krakatoa (island), 31
Kuhn, Thomas, 188
Kuiper, Gerard, 45, 72, 131, 153

L-5 Society, 206
laboratory experiments, xiv, 11, 18, 20, 36, 49, 53, 78, 98, 116, 122, 126–27, 168, 204, 211
Laplace, Marquis de, 134
Lewis, C. S., 89, 91
libration, 66
life: evolution of, xiv–xvi, 19, 62, 92, 193; extinctions of, xiv–xvi, 27, 29, 32, 88; extraterrestrial, 34, 64–65, 90, 94, 110, 165–68; origin of, 90, 105, 107; planetary effects of, 34, 106, 193; sustenance of, 33, 76–77, 81, 89, 105, 193, 201; *see also* fossils
light. *See* electromagnetic radiation (visible)
Little Prince, The (Saint-Exupéry), 44–45, 50, 52
Lowell, Percival, 165
Lucchitta, Baerbel, 182
Luna (spacecraft), vii–ix
Lunar and Planetary Institute, 112–13, 116, 122–23
Lunar Orbiter, vii, 116
Lyell, Charles, 19, 29

magnetism, planetary, 76, 83–84, 87; *see also individual planets*
man-in-space, 32, 114–15, 166, 206, 211
Mariner 2, vii
Mariner 4, vii, 17, 24, 72, 74, 165–66, 169–72
Mariner 5, vii
Mariner 6 and 7, viii, 171–72
Mariner 8, 172
Mariner 9, viii, 6, 23, 37, 73–74, 171–74, 189, 199
Mariner 10, ix, 17, 33, 37, 63–64, 70, 72–75, 82–84, 91, 100–101, *ill. 3*
Mark, Hans, 209–210
Mars, 7, 9, 23–26, 102, 123, 154, 164–183, 189–191, 194–95; *ills. 34–46*; aeolian processes, 23, 169, 178–82; atmosphere, 2, 34–35, 90, 98, 104–5, 168, 170, 173–74; atmospheric origin and evolution, 25–26, 34, 105–6, 109, 181, 195; axial tilt, 194–95; barometric pressure, 34, 72, 175, 193, 195; "canals," 64, 144, 165, 170, 173; canyons, 173, 189–91; channels, 25, 109, 172–75, 179, 181–83, 190, 195; chronology, 25, 38, 170, 174; climate and climate change, 25–26, 34–35, 106, 109–110, 166, 174, 181–83, 193–95; color, 109, 169; crater shapes, 23–24, 179; crust, 177, 179, 190–91; dust storms, 98, 169, 172–73, 176, 180, 195; erosive epoch, 24–26, 35, 109; erosive processes, 23, 166, 170, 175, 178–81, 183, 190; formation of, 59, 128; glaciation hypothesis, 174, 182–83; global topography, 189–91, 195; gravity, 98; impact cratering, 17, 23–25, 37–39, 165, 169, 180–81; interior, 25, 102, 190–91; lack of magnetic field, 84; late heavy

Mars (*continued*)
bombardment, 37–38; life on, 7, 34–35, 64–65, 110, 165–66, 168–69, 171; mission planning for, 6–7, 166, 168, 170–71, 173, 203; permafrost and ground-ice, 25–26, 105, 109, 174, 178–79, 181–83; polar regions and ice caps, 25, 34, 105, 165, 169, 174–75, 180–81, 183, 193–95; polar-wandering hypothesis, 191, 195; sapping processes, 25, 182; satellites, 165; spin, 169; tectonics, 189–91; telescopic studies, 169–70, 172; temperature, 175–76, 179; thermal evolution, 25–26, 180, 190–91; Viking lander sites, 164–67, 171, 179; volcanism, 25–26, 169, 173, 175–78, 189–91; *see also* Phobos (Martian satellite); Viking project
Mars 4 and 5 (spacecraft), ix
Maxwell, James Clerk, 10, 158–59
McCord, Tom, 52, 54
McLaughlin, Dean, 169
Mercury, 7, 63–75, 83–87, *ills. 4, 5*; chronology, 38–39, 86–87; composition, 71, 83, 85–87, 138; contraction of, 33, 86–87; density, 71, 83; drawings and maps, 63, 65, 70; formation of, 57, 83; geology, 85–86; gravity, 85, 194; impact cratering, 17, 37–39, 82, 85–86; interior, 83–87; lack of atmosphere, 64, 66–68; late heavy bombardment, 38, 86; magnetic field, 83–87; Mariner 10 planning and encounter, 72–75; misconceptions about, 64–68; radar observations, 69–72; scarps on, 33, 86; Schiaparelli's study of, 64–68; size, 64, 71; solar wind interaction, 72–73, 84; spin, 64–71, 84; temperature, 64, 68–69, 92; thermal evolution, 83, 86–87; visibility of, 63, 67–68, 75; volcanism, 85, 175, 177
Meteor Crater (Arizona), xv, 13–17, 23, 31, 36, 169
meteorites, 11–12, 15, 20, 22, 46, 51, 61, 87, 103–105, 112, 115, 123, 204; achondritic, 53; ages, 60; asteroids as sources of, 51, 53–56, 127–28; carbonaceous, 53, 57, 175; nickel-iron, 31, 47, 54; ordinary chondritic, 53–57, 83, 118; presolar system remnants in, 47, 127; stony-iron, 54–56
meteoriticists, 11–12
meteoroids, 22, 43, 161
meteorologists, 7, 92, 98–99
meteorology, dynamical. *See* atmospheres, motions and winds of
meteors, 22
metric units, 47
Mimas (Saturn satellite), 155–56, *ills. 24, 25*
minerals, 4, 51–52, 54–57, 77–78, 83, 89, 103, 107, 117–122, 124–25, 139, 145, 176, 185
minor planets. *See* asteroids and asteroid belt
moon, the, 7, 20–22, 111–128, *ills. 8–12*; Apollo exploration of, 111–17; basaltic lava origins, 119–23; chronology, 20–22, 116–17, 122; composition, 118–19, 124–26; crust, 117, 120–22; erosion on, 23; europium anomaly, 119–20; global shape, 190; gravity, 119, 121, 133; highlands, 85, 120; impact cratering, 17–18, 20–23, 36–38, 60, 85, 112,

116–17; interior, 82, 117, 119, 121–123, 134; "KREEP" rocks, 123; magma ocean model, 120–22; maria and basins, 21, 35, 85, 116, 120–23; mascons, 121; origin and formation of, 22, 37, 116, 118–19, 124–28; solar wind interaction, 84, 87, 117; thermal evolution, 47, 60, 121–22; tidal spindown, 134; volcanism, 18, 21, 23, 85, 116, 119–23, 176–77
moonquakes, 82, 115, 120–22, 134
moon rocks, 7, 12, 20–21, 48, 112, 116–121, 128, 168, 204; ages, 22, 36, 60, 115, 117, 122; compositions, 52, 117–21, 124, 127
Morrison, David, 130
Mount Palomar observatory, 44–45, 52
Mount St. Helens, 31, 135, 177, 211
Mount Wilson observatory, 149–50
Murray, Bruce, xii, 35, 37–38, 63, 74, 170–72
Mutch, Thomas A., 164, 167–68

National Academy of Sciences, 200, 202–6, 208
National Aeronautics and Space Administration (NASA), 46, 72, 113–14, 123, 170–71, 196, 198–200, 202–5, 207–9, 211–12; *see also* planetary exploration program
Neptune, x, xvi, 35

O'Leary, Brian, 198
Olympus Mons (Martian volcano), 154, 173, 175, *ills. 37–39*
Öpik, Ernst, 24, 169
orbital motion. *See* celestial mechanics
Ovenden, Michael, 48
Owen, Tobias, 72
ozone layer, xv, 34, 77, 97, 107

paleontologists, xiv–xvi
Peale, Stanton, 133–36
Phobos (Martian satellite), *ill. 47*
photodissociation, 106–9, 154
photography, telescopic, 6, 21, 44–45, 70, 100, 116
photometry, 8, 45, 73–74, 130–31
photosynthesis, 107
physicists, 3, 4, 10–11, 133, 202, 210
physics, laws of, 28, 32, 41, 124, 191
Pieri, David, 182–83
Pioneer 10 and 11, ix, 45, 59, 133, 151–52, 159, 201
Pioneer Venus, x, 84, 91, 95–97, 99, 101–102, 104, 108, 192
planetary exploration program (NASA), xvii, 112–16, 168, 192, 196–212; benefits from, 1–2, 54, 97–99, 113–14, 123, 176, 200, 206; budget and politics of, xi–xiii, 112–16, 123, 143, 168, 192, 196, 198–212; proposed future missions, 115, 168, 192, 198–99, 207, 209; public interest in, 1, 115, 166, 198, 202
Planetary Society, The, xii
planetesimals, 35–37, 57–59, 62, 103–105, 126, 139, 157
planets: interiors of, 33, 76–88; origin and formation of, 35–36, 47, 57–59, 62, 79, 103–105, 125–28, 139, 157; thermal evolution, 60–61, 79–81, 87, 121–22, 134–35, 139–40, 145, 156, 190; *see also* atmospheres; magnetism
plankton, xv–xvi, 107, 194

plate tectonics, 18, 33, 77, 107, 141–42, 176–77, 185–92, 195; *see also* continental drift
Pluto, 64
polarimetry, 68, 94
Pollack, James, 94, 108
Powell, John Wesley, 31

radar astronomy, 64, 69–72, 91, 94, 106, 159
radioactive decay, 19–20, 25–26, 33, 60–61, 80–81, 86–87, 104, 123, 139, 145
radio and radar waves. *See* electromagnetic radiation (radio)
radio astronomy, 68–69, 91, 130
Ranger spacecraft, vii, 116
Ransford, Gary, 145
Reagan, President Ronald, 192, 198–99, 208–9
religion and science, 28–29, 40
remote sensing, 51, 168; *see also* imaging from spacecraft; photography, telescopic; photometry; polarimetry; spacecraft instruments; spectrophotometry; spectroscopy
resonances, 58, 71, 134, 138, 156, 159–60, 194–95
Reynolds, Ray, 133–36
Rhea (Saturn satellite), 150, 156
Russell, C. T., 76

Sagan, Carl, xii, 1, 34–35, 92–97, 151, 153, 182
Saint-Exupéry, Antoine de, 44
satellites (artificial), 5, 10, 51, 101
satellites (natural). *See individual satellites and planets*
Saturn, 148, *ills. 26, 30*; clouds, 149; magnetic field, 76, 154, 160; radio noise, 156, 160; Voyager 1 encounter, 150–62; Voyager 2 encounter, xi–xiii, 162
Saturn's rings, 39, 148, 158–63, 210, *ills. 30–33*; E-ring and Enceladus, 156; F-ring, 157, 160–62; icy composition, 156, 158–59; particle sizes, 159–60, 162–63; relationships to moonlets, 158, 160–63; ringlets and gaps, 152, 158–63; spokes, 152, 158, 160, 162; telescopic appearance of, 148–49, 158; Voyager discoveries, 158, 160–62
Saturn's satellites, 10, 38–39, 80, 179, 192, *ill. 22*; co-orbitals, 157; geology of, 155–57; shepherding, 158, 161; Voyager 1 discoveries, 152–58; *see also individual satellites*
Schiaparelli, G. V., 64–68, 74
Schmitt, Harrison H., *ill. 12*
Schroeter, Johann, 65–66
Schultz, Peter, 191, 195
science fiction, 44, 76, 89, 91, 114, 165, 201
scientific meetings, xv–xvi, 7–8, 37, 46, 72–74, 94–95, 122–23, 204, 210–11
scientific methods, xiii–xv, 2–3, 6–12, 30, 40–42, 51, 63, 66, 94, 115–17, 133, 150–51, 153, 167, 188–89, 203, 210, 212
scientific revolutions, 17–18, 42, 184–85, 187–89
sea-floor spreading. *See* plate tectonics; continental drift
seismology. *See* earthquakes, moonquakes
Sharp, Robert, 182
Shoemaker, Eugene, 13, 31–32, 39
Sill, Godfrey, 94, 96–97
Silver, Lee, xvi
Sinton, William, 169
Smith, Bradford, xii, 129, 152–53

Soderblom, Larry, 37–38, 144
solar nebula, 46–47, 55, 57–60, 80, 83, 102, 118–19, 125, 127–28
solar system, origin of, 35, 46, 57, 61–62, 127, 201
solar wind, 33, 61, 73, 84, 91, 104, 117
Solomon, Sean, 81, 86–87
Soviet Union space program, vii–x, xi, 82, 91–92, 95, 166, 210
spacecraft. *See names of separate missions*
spacecraft instruments, 72–74, 83–84, 91, 95–96, 108, 136, 150, 162, 165–67, 192, 197, 203, 205; *see also* imaging from spacecraft
Space Shuttle, 114–15, 197–98, 207–208; Centaur upper stage, 208–209
Space Station, proposed, 115, 209
spectrophotometry, 51–53, 56, 85, 122, 127, 132, 140, 159
spectroscopy, 5, 93–94, 101, 129, 131, 152–53, 158, 169
Stone, Edward, xii
Structure of Scientific Revolutions, The (Kuhn), 188
sulfur dioxide, 95–96, 102, 130, 137
sulfuric acid, 90, 94–96, 101, 106, 108
sun, the: composition of, 53, 103; energy from, 26, 32–34, 60, 80–81, 90, 102; evolution of, 61, 81, 108, 195; origin of, 57; temperature, 92
supernovae, 47, 57, 61–62, 127
Surveyor spacecraft, viii, 116

telescopes, 5, 8, 18, 22, 31, 45, 52, 65, 71, 92, 136, 148–49, 152, 172, 212; *see also* photography
Tethys (Saturn satellite), 155–56, *ill. 27*
thermal radiation. *See* electromagnetic radiation (infrared)
tidal forces, 33, 36, 64, 71, 87, 126, 133–35, 145, 156
Titan (Saturn satellite), 152–55, *ill. 23*; atmospheric composition, 152–54; radio probing of, 153; seasons, 154–55; surface conditions, 153–54; upper atmospheric smog, 153–54

ultraviolet rays. *See* electromagnetic radiation
uniformitarianism, 28–32, 34–40, 42, 183
U.S. Congress, 113, 196, 198, 200, 202, 206–7, 212
U.S. Geological Survey (U.S.G.S.), 20, 37, 211
Uranus: formation of, 35; 1986 Voyager encounter, x, 201–2; rings, 148, 161
Urey, Harold, 12

Valles Marineris (Martian canyon), 173, 190–91, *ill. 42*
Velikovsky, Immanuel, 28, 40–42
Venera program (Soviet spacecraft), ix–x, 82, 91–92, 95, 166, *ill. 6*
Venus, 43, 89–102, 104–109, 124, 191–92, 199, *ills. 6, 7, 42*; atmosphere, 2, 90–110; atmospheric origin and evolution, 105–109; climate change, 108–109; clouds, 90–97, 100–102, 106; color, 90, 97, 102; crust and interior, 90, 102, 106–108, 192; formation of, 36, 104; geology, 91, 106, 192; impact cratering, 35; lightning, 96–97; magnetic field, 84, 87, 91; Mariner missions to, 91, 100; orbital stability

Venus (*continued*)
of, 41; phases, 90; pictures of surface, 82, 91, 95; radar probing, 72, 91, 192; size, density, and gravity of, 90, 98–99, 194; spin, 91, 98–100; tectonics, 107, 191–92; temperature and pressure of atmosphere, 90–97, 99, 107–108; topography, 91, 192; water-ice cloud model, 93–97; water on, 93, 95, 107–108; weather, 97, 99, 106; winds and atmospheric circulation, 91, 97–101; *see also* Pioneer Venus
Vesta (asteroid), 53–56
Viking project, ix, 7, 102, 164–68, 171, 174, 176, 179, 182–83, 189, 197; biology experiments, 165–66
Vishniac, Wolf, 167
volcanism, 53, 123, 132, 176–78; *see also individual planets*
Voyager project, 147, 197–98, 200, 202, 206; Voyager 1, x, xiii, 17, 38, 129–30, 132–33, 135–36, 138, 143, 150–62; Voyager 2, x, xi–xiii, 136, 143–44, 146, 155–56, 160, 162, 201

Ward, William, 195
Wasserburg, Gerald, 35–36, 204–6
wave motions, 100–101; *see also* atmospheres, motions and winds of; electromagnetic radiation
weathering (of rocks), 23, 34, 90, 106–7, 179–180
weather prediction, 2, 4, 98
Wegener, Alfred, 185–88, 194
Weidenschilling, Stuart, 50
Wetherill, George, 35–36, 39, 104
Whewell, William, 30
Worlds in Collision (Velikovsky), 40

yardangs, 180–81, *ills. 1, 46*
Young, Andrew, 74, 94